Einstein's General Theory of Relativity

Einstein's general theory of relativity can be a notoriously difficult subject for anyone approaching it for the first time, with arcane mathematical concepts such as connection coefficients and tensors adorned with a forest of indices. This book is an elementary introduction to Einstein's theory and the physics of curved space-times that avoids these complications as much as possible. Its first half describes the physics of black holes, gravitational waves, and the expanding Universe, without using tensors. Only in the second half are Einstein's field equations derived and used to explain the dynamical evolution of the early Universe and the creation of the first elements. Each chapter concludes with problem sets, and technical mathematical details are given in the appendices. This short text assumes a familiarity with special relativity and advanced mechanics.

BRIAN P. DOLAN is an Emeritus Professor in the Department of Theoretical Physics at Maynooth University in Ireland, where he taught courses in theoretical physics for 35 years, and an Adjunct Professor in the School of Theoretical Physics, Dublin Institute for Advanced Studies.

Einstein's General Theory of Relativity
A Concise Introduction

BRIAN P. DOLAN
National University of Ireland Maynooth
and
Dublin Institute for Advanced Studies

CAMBRIDGE
UNIVERSITY PRESS

Shaftesbury Road, Cambridge CB2 8EA, United Kingdom

One Liberty Plaza, 20th Floor, New York, NY 10006, USA

477 Williamstown Road, Port Melbourne, VIC 3207, Australia

314–321, 3rd Floor, Plot 3, Splendor Forum, Jasola District Centre,
New Delhi – 110025, India

103 Penang Road, #05–06/07, Visioncrest Commercial, Singapore 238467

Cambridge University Press is part of Cambridge University Press & Assessment, a
department of the University of Cambridge.

We share the University's mission to contribute to society through the pursuit of
education, learning and research at the highest international levels of excellence.

www.cambridge.org
Information on this title: www.cambridge.org/9781009263702
DOI: 10.1017/9781009263689

First published 2023

A catalogue record for this publication is available from the British Library.

ISBN 978-1-009-26370-2 Hardback
ISBN 978-1-009-26373-3 Paperback

To Mary Mulvihill,
the best science communicator I have ever known.

Contents

Preface

This book is based on lecture notes for an undergraduate course on general relativity taught at Maynooth University in Ireland over a number of years. General relativity is a notoriously difficult subject, famous for a forest of indices that obscure the underlying physics, making it difficult to extract what are often counter-intuitive phenomena such as the bending of light and the physics of black holes. In this book tensors are avoided for as long as possible; the only prerequisites are a good understanding of Lagrangian mechanics and an introductory course on special relativity, though some elementary electrodynamics, thermodynamics, and elementary quantum mechanics would be useful.

Many fascinating results from general relativity, such as the precession of the perihelion of Mercury and the bending of light by the Sun, can easily be understood without the full mathematical machinery of differential geometry. One merely needs to grasp the concept of a geodesic, the shortest path between two points in a given geometry, and learn how to determine the geodesics associated with that geometry, without needing to ask where the geometry comes from. Anyone who understands Lagrange's variational approach to classical mechanics can do this using Lagrangians quadratic in generalised coordinate velocities without the need to introduce tensors and Christoffel symbols, and this is the approach adopted here. Exceptions to this are the analytic form of the cosmic scale factor in Robertson–Walker metrics and the production of gravitational waves: the former is postponed here till after the introduction of Einstein's equations, and the latter is not treated at all, being too advanced for the level adopted here, though the propagation of gravitational waves through empty space is discussed.

After a brief introduction to the equivalence principle and the notion that the gravitational force is due to the curved geometry of 4-dimensional space-time, the first half of the book examines geodesics in various space-times before tensors are introduced. The line elements for various geometries are just written down, without justification, and the geodesics studied. The philosophy is similar to the way most students learn electromagnetism; the Coulomb field of static point charges and the magnetic field of solenoids and long, straight wires carrying electric currents are usually studied before encountering Maxwell's equations in all their glory. Here students are introduced to the Schwarzschild geometry and general properties of Robertson–Walker line elements before being faced with the full complication of Einstein's equations.

There is a distinct change of gear in the second half of the book where the mathematical description of curved geometries is developed, parallel transport is discussed, connections and Christoffel symbols are introduced, and the Riemann tensor is defined as a prelude to deriving Einstein's equations. Only then can cosmology and the Big Bang be described properly, as the dynamics of the early Universe depend crucially on the equations of state for matter and radiation through Einstein's equations.

In Chapter 4 the mathematics necessary for describing higher-dimensional curved spaces is developed. This chapter is unavoidably more abstract than the other chapters in the book, and the main results necessary for understanding Chapters 5, 6, and 7 are summarised on page 116. Chapter 4 can be omitted on a first reading except for the equations on page 116.

With a view to making the text as readable as possible, properties of the Riemann tensor are stated without proof in the body of the text, with technical details of the proofs being relegated to an appendix, as is the derivation of the Riemann tensor for Schwarzschild and Robertson–Walker space-times. The energy-momentum tensor for a relativistic fluid is also explained in a separate appendix.

It is a pleasure to thank Charles Nash and Sally Lindsay for a careful reading of the manuscript and useful suggestions, though any errors or omissions are entirely the author's own responsibility.

1

Introduction

1.1 Newtonian Gravity

No student of physics can fail to notice the similarity between Coulomb's law for the electrostatic force between two point charges q_1 and q_2 a distance r apart,

$$\mathbf{F} = \frac{1}{4\pi\epsilon_0}\frac{q_1 q_2}{r^2}\hat{\mathbf{r}}, \tag{1.1}$$

and the gravitational force between two point masses m_1 and m_2 a distance r apart, according to Newton's universal law of gravitation,

$$\mathbf{F} = -\frac{Gm_1 m_2}{r^2}\hat{\mathbf{r}}. \tag{1.2}$$

($\hat{\mathbf{r}} = \mathbf{r}/r$ is a unit vector pointing from particle 1 to particle 2, and these are the forces on particle 2 due to particle 1.) Newton published his universal law of gravitation in 1686 (though Robert Hooke appreciated the significance of the inverse square law for planetary motion before Newton and accused Newton of plagiarism), and Coulomb published his law almost exactly 100 years later, in 1785. Coulomb must have been very excited by his discovery; Equations (1.1) and (1.2) appear to imply a very close connection between electrostatics and gravity.

Even in the static case there are obvious differences, of course; the prefactor $\frac{1}{4\pi\epsilon_0}$ in Coulomb's law[1] is replaced with $-G$ in Newton's law, but this is partly just a choice of units. Instead of measuring electric charge in coulombs (the MKSA unit of charge), we could use a different set of units with $\tilde{q} = \frac{q}{\sqrt{4\pi\epsilon_0}}$ and then Coulomb's law reads

$$\mathbf{F} = \frac{\tilde{q}_1 \tilde{q}_2}{r^2}\hat{\mathbf{r}}, \tag{1.3}$$

---

[1] ϵ_0 goes by the unfortunate name of 'the electric permittivity of the vacuum'.

and instead of measuring mass in kilograms we could define $\tilde{m} = \sqrt{G}\, m$, in which case Newton's law of gravitation reads

$$\mathbf{F} = -\frac{\tilde{m}_1 \tilde{m}_2}{r^2}\hat{\mathbf{r}}. \tag{1.4}$$

($\tilde{q}$ and $\tilde{m}$ have the same units, kilograms$^{1/2}\times$ metres$^{3/2}$/second.) They look even more similar – apart from that minus sign. This mathematical similarity, however, hides the physical fact that the forces are of very different magnitudes. In MKSA units Newton's universal constant of gravitation is $G = 6.67 \times 10^{-11}$ kg^{-1}m^{3}s^{-2} while the electric permittivity of the vacuum is $\epsilon_0 = 8.85 \times 10^{-12}$C^{2}m^{3}s^{-2}kg^{-1}. The relative strengths of the gravitational to the electrostatic force between two electrons, with $q_1 = q_2 = -1.6 \times 10^{-19}$C and $m_1 = m_2 = 9.1 \times 10^{-31}$kg, is $4\pi\epsilon_0 G \left(\frac{9.1\times10^{-31}}{1.6\times10^{-19}}\right)^2 \approx 10^{-43}$, a dimensionless number which is independent of the units used. Gravity is an extremely weak force, which is why a small magnet can beat the gravitational attraction of the entire Earth and lift a metal pin off a table. Nevertheless, gravity dominates the Universe on large scales, as there is only one sign for the gravitational 'charge', m, while it is a fact from experimental observation that electric charge q can be either positive or negative, like charges repel and unlike charges attract, while m always seems to be positive and, because of that minus sign, all masses attract under Newton's gravitational force. With an equal number of positive and negative charges in the Universe, electrostatic forces cancel out in the large, while gravitational forces are cumulative and dominate on astrophysical scales.

Nevertheless, the mathematical similarity between (1.3) and (1.4) makes it very tempting to think there must be some deep relation between electromagnetism and gravity that remains to be uncovered, if only we could see a little more deeply into the nature of the two forces. Indeed, Einstein himself spent much of his later career trying to find a unified mathematical framework: a unified theory of electricity, magnetism, and gravity. He failed to achieve his dream of a unified theory. It happens that this connection is largely illusory when we start to consider moving charges and masses, particularly with velocities approaching the speed of light: we need to consider time-varying electromagnetic and gravitational fields, and the dynamics of these two fields are very different. (Einstein was well aware of this; he was motivated by deeper considerations.) Coulomb's law and magnetostatics generalise to Maxwell's equations (1865) which unify electricity and magnetism into a single mathematical framework, the theory of electromagnetism, while

Newton's law generalises to Einstein's general theory of relativity (1916), and Einstein's equations are very different from Maxwell's equations. Both predict oscillating waves, electromagnetic waves for electromagnetism (radio waves were discovered by Hertz in 1888) and gravitational waves for general relativity (discovered more recently in 2016); but the physics, and the mathematics, of the two theories turns out to be very different – that minus sign, and the absence of negative masses, are just the tip of the iceberg.[2]

It was Michael Faraday (1791–1867) who abstracted the notion of a *field* from Coulomb's law. He suggested that an electrically charged particle generates an electric field,

$$\mathbf{E} = \frac{1}{4\pi\epsilon_0} \frac{q_1}{r^2} \hat{\mathbf{r}},$$

and a second particle, with charge q, then experiences a force,

$$\mathbf{F} = q\mathbf{E},$$

in the presence of that field. Since the electrostatic force is conservative, we can rephrase this in terms of the electrostatic potential. The electric field $\mathbf{E}$ is the electric force per unit charge, and we define the electrostatic potential φ for a distribution of charges through

$$\mathbf{E} = -\nabla\varphi.$$

If we have N charges $q_1, \ldots, q_N$ at points $\mathbf{r}_1, \ldots, \mathbf{r}_N$, they generate an electrostatic potential at $\mathbf{r}$ given by

$$\varphi(\mathbf{r}) = \frac{1}{4\pi\epsilon_0} \sum_{i=1}^{N} \frac{q_i}{|\mathbf{r} - \mathbf{r}_i|},$$

and the electrostatic force experienced by a charge q at the point $\mathbf{r}$ is

$$\mathbf{F} = -q\nabla\varphi(\mathbf{r}).$$

The electric field satisfies Gauss' law and, when expressed as a differential equation, this is

$$\nabla.\mathbf{E} = -\nabla^2\varphi = \frac{\rho}{\epsilon_0}, \tag{1.5}$$

where ρ is the charge density.

---

[2] When quantum mechanics is taken into consideration, the situation is even worse: a relativistic quantum theory of electromagnetism, quantum electrodynamics, or QED, for which Richard Feynman, Sin-Itro Tomonaga, and Julian Schwinger received the Nobel Prize in Physics, was fully developed in the 1940s, but a fully credible quantum theory of gravity still eludes us.

Similar language can be used for Newtonian gravity – think of mass m_1 as creating a gravitational field,

$$\mathbf{g} = -\frac{Gm_1}{r^2}\hat{\mathbf{r}}, \tag{1.6}$$

and then another mass m experiences a force,

$$\mathbf{F} = m\mathbf{g}.$$

The gravitational force is also conservative, and $\mathbf{g}$, the acceleration due to gravity, can be defined in terms of the gravitational potential, Φ, due to a distribution of masses:

$$\mathbf{g} = -\nabla\Phi. \tag{1.7}$$

If we have N point masses $m_1, \ldots, m_N$ at points $\mathbf{r}_1, \ldots, \mathbf{r}_N$, they generate the gravitational potential

$$\Phi(\mathbf{r}) = -G\sum_{i=1}^{N}\frac{m_i}{|\mathbf{r} - \mathbf{r}_i|}, \tag{1.8}$$

and the gravitational force on a mass m at the point $\mathbf{r}$ is

$$\mathbf{F} = -m\nabla\Phi(\mathbf{r}) = m\mathbf{g}. \tag{1.9}$$

The gravitational field $\mathbf{g}$ satisfies the gravitational version of Gauss' law, Poisson's equation,

$$\nabla.\mathbf{g} = -\nabla^2\Phi = -4\pi G\rho, \tag{1.10}$$

where ρ is the mass density. Again, the similarity between (1.5) and (1.10) can be traced to the fact that the Coulomb force and the Newtonian gravitational force are both inverse square laws; the only difference is that $\frac{\rho}{\epsilon_0}$ in (1.5) is replaced with $-4\pi G\rho$ in (1.10).

Forces cause things to accelerate, and here we meet an important difference between electricity and gravity (which will turn out to be crucial): mass plays two completely different roles in Newton's universal law of gravitation and in Newton's second law. In Newton's universal law of gravitation (1.2), the gravitational field is (1.6) and m_1 is like a gravitational charge; it is a 'charge' generating a gravitational field. On the other hand, in Newton's second law, (1.9) written as

$$\mathbf{F} = m_I\mathbf{a},$$

with $\mathbf{a} = \mathbf{g}$, m_I is the inertial mass, a measure of the reluctance, or inertia, of a body to be accelerated. Two bodies with different inertial masses experiencing the same force will undergo different accelerations; the one with the larger inertial mass will accelerate more slowly.

For example, a proton and a singly ionised atom of helium $^4\text{He}^+$ both experience the same force in a static electric field $\mathbf{E}$,

$$\mathbf{F} = q\mathbf{E},$$

where $q = 1.60 \times 10^{-19}\text{C}$ is the charge of a proton. But they will consequently have different accelerations – for the proton,

$$\mathbf{a} = \frac{q}{m_I}\mathbf{E},$$

where $m_I = 1.67 \times 10^{-27}\text{kg}$ is the inertial mass of a proton, while for the ionised helium,

$$\mathbf{a} = \frac{q}{M_I}\mathbf{E},$$

where M_I is the inertial mass of a helium atom (to a reasonable approximation some four times m_I). Under the *same* force, the helium atom undergoes an acceleration some four times *less* than that of the proton.

However, put the proton and the helium atom in the same gravitational field $\mathbf{g}$ and the proton will experience a force determined by Newton's Universal Law of Gravitation (1.2), analogous to $\mathbf{F} = q\mathbf{E}$ in electrostatics,

$$\mathbf{F} = m_G\,\mathbf{g},$$

where m_G here is the gravitational mass of the proton, while the helium atom will feel a different force,

$$\mathbf{F} = M_G\,\mathbf{g},$$

where M_G is the gravitational mass of the helium atom.

The proton then experiences an acceleration,

$$\mathbf{a} = \frac{\mathbf{F}}{m_I} = \frac{m_G}{m_I}\mathbf{g},$$

while for the helium atom,

$$\mathbf{a} = \frac{\mathbf{F}}{M_I} = \frac{M_G}{M_I}\mathbf{g}.$$

Since inertial mass equals gravitational mass,[3] $m_I = m_G$ and $M_I = M_G$, both the proton and the helium undergo *exactly* the same acceleration in the gravitational field,

$$\mathbf{a} = \mathbf{g}.$$

---

[3] More generally, inertial mass and gravitational mass are proportional to one another, $m_I = km_G$, with k some constant. But we are free to re-scale m_G by changing the units in which we define Newton's constant, and it is a convention, albeit a very natural one, to define G so that $k = 1$ and $m_I = m_G$. From now on we shall not distinguish between gravitational mass and inertial mass.

The equality of gravitational and inertial mass is related to the difference between weight and mass. In elementary physics courses mass is usually defined as force divided by acceleration,

$$m = \frac{F}{a}.$$

This is really the inertial mass. Weight, on the other hand, is the force you feel on your feet when you stand on the floor, as a result of the acceleration due to gravity; the upward force is

$$F = -mg,$$

if g is downwards, and m here is the gravitational mass. In outer space, where $g = 0$, you are weightless, not massless.

This aspect of the gravitational force was understood even before Newton discovered his Universal Law of Gravitation. There is the tale (probably apocryphal) of Galileo dropping two different weights from the top of the leaning tower of Pisa. A more modern demonstration was when the Apollo 15 commander David Scott dropped a geological hammer and a falcon's feather onto the surface of the Moon and the world saw them accelerate downwards at exactly the same rate and hit the lunar surface at exactly the same time, despite the huge disparity in their masses.[4]

This equivalence of inertial mass and gravitational mass has very far reaching consequences. For example, it is impossible to tell the difference between a uniform constant gravitational field and a constant acceleration. Imagine standing in a featureless closed room, or box, with no windows; you feel a force on your feet from the floor pushing against you as the gravitational field of the Earth tries to accelerate you downwards, but the solid floor stops the acceleration with an equal and opposite force. Now imagine the box is in empty space far away from any source of gravity, but it is in a rocket with a thruster that is accelerating it at precisely 9.81 m s^{-2}. You cannot tell the difference between the effects of the acceleration due to the rocket's thrusters and the Earth's gravitational field – they are indistinguishable because gravitational mass equals inertial mass (see the two pictures on the right-hand side of Figure 1.1).

Now suppose the box is in a lift shaft and is suspended by a metal cable. You feel the force of the floor on your feet. If the cable snaps, however, you will suddenly be in free fall and the force on your feet

---

[4] www.youtube.com/watch?v=KDp1tiUsZw8

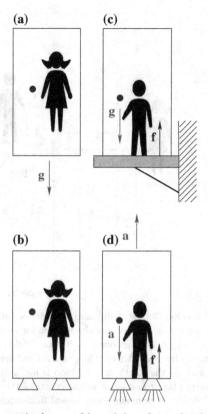

Figure 1.1 **The equivalence of inertial and gravitational mass.** The situation in a freely falling lift (a) is indistinguishable from that in a rocket in empty space far removed from any gravitational field (b); objects just float and experience no forces. An observer in a static box in a gravitational field feels a force on his feet due to his weight $\mathbf{f} = -m\mathbf{g}$ (c) which is indistinguishable from the force due to acceleration ($\mathbf{f} = -m\mathbf{a}$) of an accelerating rocket in empty space (d).

will disappear – as long as the box continues to accelerate downwards, without hitting anything, you will float freely inside the box as if there were no gravitational field at all, just like an astronaut in empty space in a rocket that is not accelerating. (See the two pictures in parts (a) and (b) of Figure 1.1.)

This is only true for uniform gravitational fields. In fact, the Earth's gravitational field is not uniform; it converges, and changes in magnitude, as one moves towards the centre of the Earth. If our experimenter

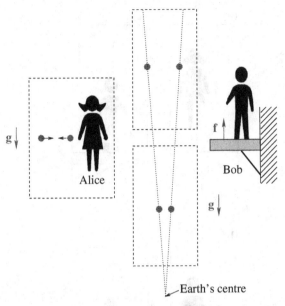

Earth's centre

Figure 1.2 **Tidal forces.** Bob, standing and watching a falling lift acceler-ate past him, feels a force on his feet opposite in direction to the acceleration due to gravity $\mathbf{f} = -m\mathbf{g}$. A non-uniform gravitational field can be detected by someone floating in free fall (Alice). The gravitational field generated by a spherical mass such as the Earth or the Moon is not uniform over large distances; it converges towards a point at the centre of the mass, and this gives rise to tidal forces. A uniform gravitational field would not generate any tidal forces.

who is freely falling in the lift shaft (Alice) takes two objects (such as a hammer and a feather) and releases them together at rest, they will just float in front of her. But if she waits long enough (about 20 minutes for free fall in a tunnel drilled through the Earth), she will see them start to drift towards each other as the box gets nearer to the centre of the Earth (see Figure 1.2). She will then know that there is a non-uniform gravitational field present – this is an example of a tidal force. However, as long as we restrict our considerations to regions of space over which a gravitational field does not vary appreciably in either magnitude or direction, we cannot tell the difference between a gravitational field and an acceleration – this is known as the *Equivalence Principle*.[5]

---

[5] Most textbooks distinguish between different kinds of Equivalence Principle, the *Weak* Equivalence Principle and the *Strong* Equivalence Principle, depending on how widely it is applied to physical phenomena. We shall not make that distinction here.

1.2 Equivalence Principle

Because inertial mass equals gravitational mass, all massive objects follow the same trajectory in a gravitational field, if they start from the same place with the same velocity. This observation led Einstein to suggest that the trajectory of a falling body is not determined by any properties of the body itself, such as its internal construction or constituent parts, but by the properties of the space (more correctly space-time) in which the body moves – and the trajectories are not straight because space-(time) is curved. In fact, freely falling bodies follow trajectories between two points in space-time that extremise the time it takes to go from one event to the other, as measured by a clock carried by the body (called the body's *proper time*).[6]

A simple example from 3-dimensional geometry illustrates Einstein's thinking. Consider an airplane flying from Delhi to Vancouver. To minimise fuel costs, the pilot wants to follow a route which takes the shortest path between the two cities – this takes her close to the North Pole. Now consider a second pilot flying from Mumbai to San Francisco; to minimise his fuel costs he takes a trajectory that takes him south, as in Figure 1.3. We see from the figure that the planes initially diverge, then their trajectories become parallel, and then they start converging. Each of the pilots is taking the shortest path to their destination, and to each pilot it looks as though the other plane is initially moving away but is being pulled inexorably towards them by some 'force' which depends not on the properties of the planes, but on the curvature of the Earth.

On a curved surface, the angles of a triangle do not necessarily add up to 180°. For example, on the surface of the Earth the angles of the triangle with one vertex at the north pole, one in Quito (the capital of Ecuador, on the equator 80°W of the Greenwich meridian), and the third in Libreville (the capital of Gabon, on the equator 10°E of the Greenwich meridian), the angles add to 90° + 90° + 90° = 270°.

The German mathematician Gauss, one of the fathers of non-Euclidean geometry, was involved in a land survey in Germany and made many trigonometric measurements between 1818 and 1832. These included measuring the angles of a large triangle with vertices at the tops of three prominent hills near Göttingen in northern Germany. Within

---

[6] We shall see that, for massive bodies, the proper time is *maximised*. There is a principle in optics, *Fermat's principle of least time*, which states that the path taken by a beam of light in a refractive medium is that which *minimises* the time taken for the light to travel between two fixed points within the medium, but this is not the proper time of the light beam.

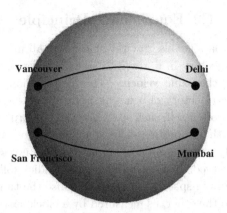

Figure 1.3 **Great circles.** Two airplanes taking paths of shortest length on the surface of the Earth. Technically the two paths are called *great circles*; they are circles whose centre is the centre of the Earth and whose radius is the radius of the Earth.

the accuracy of the measurements, the angles added up to the Euclidean value of 180°, giving no evidence of any curvature on the length scales that were accessible to him.

A modern measurement was made by the Planck satellite in 2015. This set limits on any spatial curvature of our Universe over cosmological distances: if it is non-zero, it must involve length scales at least of order 10^{28}m and possibly greater – some 100 times greater than the size of the observable Universe. The Planck results are compatible with zero spatial curvature.

However, any efforts to understand gravity in terms of the curvature of 3-dimensional space were doomed to failure. The analogy with the airline pilots is misleading, because in a gravitational field bodies with different starting velocities have different trajectories; this is an indication that in gravity time must somehow enter the picture as well as space. Having formulated his special theory of relativity in 1905 (which did not include gravity), Einstein was in a unique position to make progress in developing this idea further by postulating that, not just 3-dimensional space but *4-dimensional space-time* is curved. Einstein's interpretation of gravity is that tidal forces are a manifestation of a curvature of 4-dimensional space-time and all particles starting from the same point at the same time with the same velocity will follow the same trajectory in a curved 4-dimensional space-time. This is Einstein's Equivalence Principle at work. Einstein commented some years after developing the general theory of relativity that, when this idea occurred to him, it was 'the happiest thought of my life'.[7]

Understanding the consequences of these ideas requires studying the geometry of curved 4-dimensional space-times. This is inevitably rather technical mathematically, but before embarking on that journey let us emphasise the philosophy here. In order to understand physics in a gravitational field, we take the known laws of physics in flat 4-dimensional space-time (called *Minkowski* space-time) and try to write them out for a curved space-time. The complications of general relativity lie mainly in learning how to write the laws of physics in a curved space-time.

Problems

1) Calculate the speed of a planet moving in a circular orbit of radius r around a star of mass M. (Ignore the mass of the planet relative to M.) Show that

$$v^2 = \frac{GM}{r}.$$

2) Evaluate the escape velocity from the surface of a planet of mass M and radius R. What is its value when $R = 2GM/c^2$, where c is the speed of light?

3) **The geometry of an ellipse:** Kepler's first law of planetary motion states that the planets move around the Sun in an ellipse with the Sun at one focus. An ellipse is a very precise shape, and we will need the mathematical formulation of that shape when we discuss the Schwarzschild space-time later.

   In Cartesian coordinates (x', y'), with O' as the origin, the equation of an ellipse is

   $$\frac{x'^2}{a^2} + \frac{y'^2}{b^2} = 1.$$

   If $a > b$, a is called the semi-major axis (half the larger diameter) and b is the semi-minor axis (half the smaller diameter).

   An ellipse can be drawn on a piece of paper by tying a piece of string into a loop of length l, fixing two drawing pins into the paper a distance d apart, looping the string around the drawing pins (this requires $l > 2d$) and using the tip of a pencil to pull the string taut

---

[7] Einstein, A., *Fundamental ideas and methods of the theory of relativity, presented in their development: in Collected papers of Albert Einstein vol. 7: The Berlin years 1918–1921* (English translation), Princeton University Press (2002).

and then move the pencil in an arc around the drawing pins, keeping the string taut.

In the following unnumbered drawing, O' is the symmetric centre of the ellipse, one drawing pin is at O, and the other is at $\tilde{O}$, a distance $\Delta = \frac{d}{2}$ from O', to its left. The points O and $\tilde{O}$ are the called the *foci* of the ellipse.

Show that:

a) $l = 2(\Delta + a)$.

b) $r + \tilde{r} = 2a$.

c) Define r_0 and e as shown in the figure: the distance from O to A is $\frac{r_0}{1-e}$ and the distance from O to B is $\frac{r_0}{1+e}$, with $0 \le e < 1$. Show that:

    i) $a = \frac{r_0}{1-e^2}$;

    ii) the distance between the centre O' and a focus O is $\Delta = \frac{er_0}{1-e^2}$;

    iii) $b = \frac{r_0}{\sqrt{1-e^2}}$;

    iv) $\frac{2r_0}{1-e} = l$;

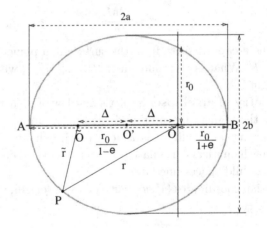

d) the equation of the ellipse in polar coordinates (r', θ') relative to O' is

$$r'^2 = \frac{r_0^2}{(1-e^2)(1-e^2\cos^2\theta')};$$

e) the equation of the ellipse in polar coordinates (r, θ) relative to the focus O is

$$r = \frac{r_0}{(1+e\cos\theta)}.$$

f) the distance between the focus $\tilde{O}$ and P is

$$\tilde{r} = \frac{r_0(1 + 2e\cos\theta + e^2)}{(1 - e^2)(1 + e\cos\theta)}.$$

g) Calculate the area enclosed by the ellipse.

Note: you may find the following integral useful:

$$\int_0^{2\pi} \frac{d\theta}{(1 + e\cos\theta)^2} = \frac{2\pi}{(1 - e^2)^{3/2}}.$$

4) Treating the Earth as a perfect sphere uniformly covered in water, calculate the height of the tides raised by the Moon (ignoring the Sun) and the height of the tides raised by the Sun (ignoring the Moon). Which is the larger effect?

(The Earth has mass 5.97×10^{24}kg and equatorial radius 6,370 km; the Moon has mass 7.35×10^{22}kg and is 384,000 km from the Earth on average; the Sun has mass 2.0×10^{30}kg and is 150 million km from the Earth on average.)

2

Metrics

A *metric* is a rule for assigning lengths to paths between two points. Think of it as a 'distance function' – it gives the distance between two points along a path between them. This distance will, of course, depend not just on the two end points but also the chosen path; but we can get a good handle on the metric by studying points that are infinitesimally close to one another. We shall start by studying some familiar geometries.

2.1 Euclidean Space

Where better to start a discussion of geometry than with Pythagoras' theorem? The square of the distance between two points A and B in the 2-dimensional plane with Cartesian coordinates (x_A, y_A) and (x_B, y_B) is

$$(\Delta s)^2 = (x_B - x_A)^2 + (y_B - y_A)^2 = \Delta x^2 + \Delta y^2,$$

where $\Delta x = x_B - x_A$ and $\Delta y = y_B - y_A$ (s for separation). Although the definition of Cartesian coordinates in Euclidean space requires choosing an origin, separations do not depend on this choice: if we shift both x_B and x_A by the same amount, Δs does not change; similarly for y_B and y_A. Also, rotating the x and y axes by some constant angle about any axis does not change Δs.

$$\begin{pmatrix} x' \\ y' \end{pmatrix} = \begin{pmatrix} \cos\phi & \sin\phi \\ -\sin\phi & \cos\phi \end{pmatrix} \begin{pmatrix} x \\ y \end{pmatrix}$$

leaves Δs invariant:

$$(\Delta s)^2 = \Delta x^2 + \Delta y^2 = (\Delta x')^2 + (\Delta y')^2.$$

For infinitesimal separations we let $\Delta x \to dx$, $\Delta y \to dy$, and

$$ds^2 = dx^2 + dy^2. \tag{2.1}$$

The infinitesimal form (2.1) is sometimes called a *line element*, really the square of the length of an infinitesimally short piece of a line.

In 3-dimensional Euclidean space, distances are given by

$$(\Delta s)^2 = (x_B - x_A)^2 + (y_B - y_A)^2 + (z_B - z_A)^2 = \Delta x^2 + \Delta y^2 + \Delta z^2.$$

For infinitesimal separations,

$$ds^2 = dx^2 + dy^2 + dz^2 = \begin{pmatrix} dx & dy & dz \end{pmatrix} \begin{pmatrix} 1 & 0 & 0 \\ 0 & 1 & 0 \\ 0 & 0 & 1 \end{pmatrix} \begin{pmatrix} dx \\ dy \\ dz \end{pmatrix}. \tag{2.2}$$

It is often convenient to adopt an index notation, $x^1 = x$, $x^2 = y$, $x^3 = z$, and write the line element as

$$ds^2 = \sum_{\alpha,\beta=1}^{3} \delta_{\alpha\beta} dx^\alpha dx^\beta = \delta_{\alpha\beta} dx^\alpha dx^\beta, \tag{2.3}$$

where $\delta_{\alpha\beta}$ with $\alpha, \beta = 1, 2, 3$ are the components of the identity matrix and the Einstein summation convention is adopted. In 2.3 we adopt the convention that summation signs are omitted, on the understanding that any index that appears twice in any term is summed over its range of values. If an index appears once in any term, then the same index must appear once in all other terms and there are three equations or, more generally, D equations in D dimensions.[1]

By integrating infinitesimal line elements along any curve, we can calculate the length L_{PQ} of a curve between two fixed points P and Q. If we parameterise the curve by λ and describe it by $(x(\lambda), y(\lambda), z(\lambda))$, then the length is

$$L_{PQ} = \int_{\lambda_P}^{\lambda_Q} \sqrt{\left(\frac{dx}{d\lambda}\right)^2 + \left(\frac{dy}{d\lambda}\right)^2 + \left(\frac{dz}{d\lambda}\right)^2} \, d\lambda.$$

Alternatively, we might wish to use spherical polar coordinates (r, θ, ϕ) to describe the geometry of 3-dimensional space,

---

[1] This notation was introduced by Einstein, who realised that simply omitting the summation signs did not usually result in any ambiguity in the formulae and made them easier to read, so it is called the *Einstein summation convention*. With some practice it becomes clear that this is not a silly thing to do. (If an index appears three times in the same term, then you have made a mistake!)

$$x = r\sin\theta\cos\phi$$
$$y = r\sin\theta\sin\phi$$
$$z = r\cos\theta,$$

in which case

$$ds^2 = dr^2 + r^2(d\theta^2 + \sin^2\theta d\phi^2).$$

In index notation with $x^{1'} = r$, $x^{2'} = \theta$, and $x^{3'} = \phi$ (the prime is to distinguish r, θ, ϕ coordinates from x, y, z coordinates in the index notation),

$$ds^2 = g_{\alpha'\beta'}dx^{\alpha'}dx^{\beta'}$$

with

$$g_{\alpha'\beta'} = \begin{pmatrix} 1 & 0 & 0 \\ 0 & r^2 & 0 \\ 0 & 0 & r^2\sin^2\theta \end{pmatrix}. \qquad (2.4)$$

The circumference of a circle, for example, is easily calculated in polar coordinates. Let the circle lie in the x-y plane and set r constant and fix $\theta = \frac{\pi}{2}$. Then we can take the parameterisation of this curve to be simply $\phi(\lambda) = 2\pi\lambda$, with $0 \le \lambda < 1$, and an infinitesimal segment of the circle has line element

$$ds^2 = (rd\phi)^2 = \left(r\frac{d\phi}{d\lambda}\right)^2 (d\lambda)^2 = (2\pi rd\lambda)^2.$$

The length is thus

$$L = \int_0^1 \sqrt{(2\pi rd\lambda)^2} = 2\pi r \int_0^1 d\lambda = 2\pi r.$$

Note that, when $\theta = 0$ (the z-axis), arbitrary changes in ϕ, with r and θ fixed, give zero length – ϕ is not a good coordinate to use on the z-axis.

In any given geometry, the *shortest* path between two fixed points is called a *geodesic*.

2.2 Alternative Metrics in 2-Dimensional Space

There can be more than one way of defining lengths in a space. Consider, for example, the picture of a receding rail track in Figure 2.1. From a flat 2-dimensional perspective, the railway sleepers nearer the horizon

Figure 2.1 An image with two different natural metrics, (2.5) and (2.6).

seem smaller, and this is captured by the usual Euclidean line element in 2-dimensional polar coordinates, with the vanishing point as the origin:

$$ds^2 = dr^2 + r^2 d\phi^2. \tag{2.5}$$

From a 3-dimensional perspective, however, we suspect that the railway sleepers are all the same length, $\tilde{l}$ say, so a 3-dimensional viewer would think that the lengths of the sleepers in the image are all identical and would be better described by

$$ds^2 = dr^2 + \tilde{l}^2 d\phi^2 = \tilde{l}^2 (d\tilde{r}^2 + d\phi^2), \tag{2.6}$$

where $\tilde{r} = r/\tilde{l}$ is dimensionless. Equations (2.5) and (2.6) are different metrics on the same space; both give perfectly valid geometries, but they are different geometries.[2]

---

[2] If we use the height of the fence posts, say $\bar{l}$, instead of the length of the sleepers $\tilde{l}$ we would get

$$ds^2 = dr^2 + \bar{l}^2 d\phi^2,$$

which is a different metric if $\bar{l} \neq l$. But it is only a matter of fixing the overall length scale. Let $\bar{r} = r/\bar{l}$; then

$$ds^2 = \bar{l}^2 (d\bar{r}^2 + d\phi^2)$$

2.3 The Sphere – the Map Maker's Problem

The equation of a sphere of radius a is

$$x^2 + y^2 + z^2 = a^2, \tag{2.7}$$

so

$$z(x,y) = \pm\sqrt{a^2 - x^2 - y^2} \qquad \Rightarrow \qquad dz = \mp\frac{xdx + ydy}{\sqrt{a^2 - x^2 - y^2}}$$

and, when restricted to the surface of the sphere, the line element is

$$ds^2 = dx^2 + dy^2 + \frac{(xdx + ydy)^2}{a^2 - x^2 - y^2}$$

$$= \frac{(a^2 - x^2 - y^2)(dx^2 + dy^2) + x^2dx^2 + y^2dy^2 + 2xydxdy}{a^2 - x^2 - y^2}$$

$$= \frac{(a^2 - y^2)dx^2 + (a^2 - x^2)dy^2 + 2xydxdy}{a^2 - x^2 - y^2}$$

$$= \begin{pmatrix} dx & dy \end{pmatrix} \begin{pmatrix} \frac{a^2-y^2}{a^2-x^2-y^2} & \frac{xy}{a^2-x^2-y^2} \\ \frac{xy}{a^2-x^2-y^2} & \frac{a^2-x^2}{a^2-x^2-y^2} \end{pmatrix} \begin{pmatrix} dx \\ dy \end{pmatrix}.$$

Here we are using x and y, with $x^2 + y^2 \le a^2$, as two coordinates labelling points on the sphere, which is a 2-dimensional curved surface. The 2×2 matrix

$$g_{\alpha\beta} = \frac{1}{(a^2 - x^2 - y^2)} \begin{pmatrix} a^2 - y^2 & xy \\ xy & a^2 - x^2 \end{pmatrix} \tag{2.8}$$

is the metric for this 2-dimensional curved space. (We can always choose $g_{\alpha\beta}$ to be symmetric in its indices $g_{\alpha\beta} = g_{\alpha\beta}$.)

If both x and y are small relative to a, $\frac{x}{a} \sim \frac{y}{a} \sim \epsilon$ with $\epsilon \ll 1$, then

$$g_{\alpha\beta} = \delta_{\alpha\beta} + O\left(\epsilon^2\right).$$

As long as we do not stray too far away from $x = y = 0$, the metric is indistinguishable from the 2-dimensional Euclidean metric (2.1). For example, we can draw maps covering regions of (10 km^2) or so of the Earth's surface and put square grid-lines on them which make things look flat; but on length scales of 100 km or more, distortions inevitably creep in. This is a generic feature of curved spaces; they look flat on small enough length scales, and this will be key to our understanding of curved space-times and gravity.

is the same as (2.6), up to an overall scale. Equation (2.6) is actually a metric on a cylinder with radius $\tilde{l}$ and r measuring length along the cylinder.

Note that the entries in the metric (2.8) diverge when $x^2+y^2 = a^2$ (i.e. on the equator, $z = 0$), although the geometry of the sphere is perfectly regular there. There is no pathology in the geometry when $x^2 + y^2 = a^2$; it is just that x and y are not good coordinates on the equator. This is a phantom singularity, called a *coordinate singularity* – a consequence of a bad choice of coordinates on the equator – and we shall frequently encounter examples of this. It is often the case that there is no one choice of coordinates that is good for every point of a curved space.

We can again choose to label our two coordinates by indices, for example, $x = x^1$, $y = x^2$, and use x^α, with $\alpha = 1, 2$, rather than x and y. Then we can write, using the summation convention,

$$ds^2 = g_{\alpha\beta}(x)dx^\alpha dx^\beta,$$

where $g_{\alpha\beta}(x)$ are the components of the matrix (2.8) and the argument x stands for the set of x^α.

The Cartesian coordinates x and y give the rather inelegant expression (2.8) for the metric on a sphere. While this can be useful near $x = y = 0$, where ϵ quantifies deviations from flat space, using this expression on larger length scales causes severe distortions (like the apparent size of Greenland on rectangular world maps – it looks much larger than India, despite the fact that the area of India is 50 per cent more than that of Greenland). Better is to use 3-dimensional polar coordinates (r, θ, ϕ) in which the 3-dimensional line element is (2.4). The sphere is then obtained by setting $r = a = const$, giving

$$ds^2 = a^2(d\theta^2 + \sin^2\theta d\phi^2). \tag{2.9}$$

This gives the metric on a 2-dimensional sphere of radius a in polar coordinates, with $x^{1'} = \theta$, $x^{2'} = \phi$, and the metric takes the form

$$g_{\alpha'\beta'} = \begin{pmatrix} a^2 & 0 \\ 0 & a^2\sin^2\theta \end{pmatrix}. \tag{2.10}$$

A sphere is, of course, a highly symmetrical shape. It is obvious from (2.9) that rotating the sphere by, for example, shifting ϕ by a constant (rotating around the z-axis) is a symmetry of the line element (2.9).

In the parameterisation (2.9), we can easily calculate the length L_θ of any circle of constant latitude, fix θ so $ds = a\sin\theta d\phi$, and integrate ϕ from 0 to 2π:

$$L_\theta = \int_0^{2\pi} \sqrt{ds^2} = \int_0^{2\pi} a\sin\theta d\phi = a\sin\theta \int_0^{2\pi} d\phi = 2\pi a\sin\theta.$$

As before, we can calculate the length L_{PQ} of any curve between two fixed points P and Q. If we parameterise the curve by λ and describe it by $(\theta(\lambda), \phi(\lambda))$, then the length is

$$L_{PQ} = a \int_{\lambda_P}^{\lambda_Q} \sqrt{\left(\frac{d\theta}{d\lambda}\right)^2 + \sin^2 \theta \left(\frac{d\phi}{d\lambda}\right)^2} \, d\lambda.$$

Yet another way of parameterising a sphere is to define a new coordinate,

$$\tilde{r} = a \sin \theta, \tag{2.11}$$

in terms of which

$$a^2 d\theta^2 = \frac{d\tilde{r}^2}{\cos^2 \theta} = \frac{d\tilde{r}^2}{1 - \frac{\tilde{r}^2}{a^2}}$$

and

$$ds^2 = \frac{d\tilde{r}^2}{1 - \frac{\tilde{r}^2}{a^2}} + \tilde{r}^2 d\phi^2,$$

with $0 \leq \tilde{r} < a$. In this form we can see immediately that the sphere is flat in the limit $a \to \infty$, since then

$$d\tilde{r}^2 + \tilde{r}^2 d\phi^2 \tag{2.12}$$

is just the flat 2-dimensional Euclidean metric in polar coordinates.

When we study cosmology later, we shall use a slight modification of this notation and define a dimensionless variable, $r = \frac{\tilde{r}}{a}$ (not the radial coordinate in 3-dimensional polar coordinates here), and write

$$ds^2 = a^2 \left(\frac{dr^2}{1 - Kr^2} + r^2 d\phi^2\right). \tag{2.13}$$

Then, $K = +1$ with $0 \leq r < 1$ is a sphere of radius a (actually only half a sphere, since $0 \leq r < 1$ corresponds to the northern hemisphere $0 \leq \theta < \frac{\pi}{2}$ in (2.11)), while $K = 0$ with $0 \leq r < \infty$ is a flat 2-dimensional plane, in which r *is* the radial coordinate in 2-dimensional polar coordinates. These coordinates are shape-shifters!

It is not essential that $K = +1$; the real distinction here is between $K = 0$ and $K > 0$. For any $K > 0$, let $r' = \sqrt{K} r$ and

$$ds^2 = \frac{a^2}{K} \left(\frac{dr'^2}{1 - r'^2} + r'^2 d\phi^2\right)$$

is the line element on a sphere of radius $\frac{a}{\sqrt{K}}$.

2.4 Flat Space-Time

In special relativity, time enters as a coordinate, and a point (an event) in space-time is uniquely fixed by specifying x, y, and z in space, as well as the time t at which the event occurred. Using the speed of light c, we can treat ct as a time coordinate with dimensions of length. (Alternatively, we could use coordinates with dimensions of time t, x/c, y/c, and z/c – for example, astronomical distances are measured in years, light-years.)

The relativistic 'distance' between two events A and B with coordinates (ct_B, x_B, y_B, z_B) and (ct_A, x_A, y_A, z_A) is obtained from

$$\Delta s^2 = \Delta x^2 + \Delta y^2 + \Delta z^2 - c^2 \Delta t^2 \tag{2.14}$$

where $\Delta t = t_B - t_A$. The geometry associated with this notion of 4-dimensional distance is known as *Minkowski space-time*.

Warning! Despite the notation, Δs^2 can be either positive or negative, depending on the relative magnitudes of Δt, Δx, Δy, and Δz.

This is a strange geometry – it is possible for the Minkowski space-time 'distance' between two different events to be zero. For example a flash of light leaving the spatial origin $x = y = z = 0$ at time $t = 0$ expands in a shell of radius

$$r = \sqrt{x^2 + y^2 + z^2} = ct$$

and every point on this shell has zero 'distance' from the initial event $(0, 0, 0, 0)$. Two events with $\Delta s^2 = 0$ are said to have *light-like* separation. If $\Delta s^2 > 0$, the events have *space-like* separation; if $\Delta s^2 < 0$, they have *time-like* separation. Two events with space-like separation cannot have any physical influence on one another – to do so would require sending information faster then the speed of light; indeed, the events' order in time can be changed by observing them from different inertial reference frames. Two events with time-like or light-like separation have a physical ordering; one always occurs before the other in all reference frames, and the earlier one can influence the later one.

The Minkowski line element is symmetric under Lorentz transformations in the same manner that the Euclidean line element is symmetric under rotations. Under a Lorentz transformation, for example, a boost with speed v in the x-direction, the transformation, together with its inverse, is

$$ct' = \gamma(v)(ct - vx/c), \qquad ct = \gamma(v)(ct' + vx'/c),$$
$$x' = \gamma(v)(x - vt), \qquad x = \gamma(v)(x' + vt'),$$
$$y' = y, \qquad y = y',$$
$$z' = z, \qquad z = z',$$

where $\gamma(v) = \frac{1}{\sqrt{1-\frac{v^2}{c^2}}}$. Thus,

$$c\Delta t = \gamma(v)(c\Delta t' + v\Delta x'/c),$$
$$\Delta x = \gamma(v)(\Delta x' + v\Delta t'),$$
$$\Delta y = \Delta y',$$
$$\Delta z = \Delta z',$$

and it is easy to check that

$$ds^2 = \Delta x^2 + \Delta y^2 + \Delta z^2 - c^2\Delta t^2 = (\Delta x')^2 + (\Delta y')^2 + (\Delta z')^2 - c^2(\Delta t')^2$$

is unchanged when we transform between the two inertial coordinate systems.

For two events with time-like separation, $\Delta s^2 < 0$ and $\sqrt{-\Delta s^2}/c$ has an important physical interpretation. Consider an observer, a man (Bob) wearing a Rolex™ watch, for example,[3] at rest at the spatial origin in the primed coordinate system just presented. Then, two events in Bob's history could be a 'tick' at time t'_1 and a 'tock' at time t'_2, with $t'_2 > t'_1$ and $\Delta t' = t'_2 - t'_1$. The tick and the tock have a time-like separation with

$$\Delta s^2 = -c^2(\Delta t')^2,$$

and the time between these two events is called the *proper time*, $\Delta\tau = \Delta t' = \sqrt{-\Delta s^2}/c$, the time measured in the observer's own reference frame. Thus, Bob's proper time between the tick and the tock is

$$\Delta\tau_B = \Delta t'.$$

However, in the unprimed reference frame the watch is moving with speed v, and the time difference between the two events is

$$\Delta t = \gamma(v)(\Delta t' + v\Delta x'/c^2) = \gamma(v)\Delta t' = \gamma(v)\Delta\tau_B,$$

since the watch is not moving in the primed reference frame and $\Delta x' = 0$. So, as the watch moves past another observer (Alice) at rest in the

---

[3] This 'observer' could be a man with an expensive watch or it could simply be a particle that is unstable under nuclear decay, such as a neutron which decays to a proton, an electron, and an anti-neutrino with a half-life of 10 minutes. An unstable neutron carries its own clock within it.

unprimed reference frame, on her watch the time between the tick and the tock is

$$\Delta t = \gamma(v)\Delta\tau_B > \Delta\tau_B.$$

This is the famous special relativistic time-dilation effect. Δt is not Lorentz invariant, but the proper time of the watch,

$$\Delta\tau_B = \frac{\sqrt{|\Delta s^2|}}{c}, \tag{2.15}$$

is.

Another way of saying this is that

$$\Delta\tau_B = \Delta t'$$

is the proper time shown by Bob's watch between the tick and the tock, while

$$\Delta\tau_A = \Delta t$$

is the time interval shown by a clock carried by Alice, in whose reference frame the Bob's watch is moving with speed v.

$$\Delta\tau_A > \Delta\tau_B$$

and, according to Alice, Bob's watch is running slow.[4]

An important lesson to learn from this is that the distance $\sqrt{|\Delta s^2|}$ along a time-like trajectory is c times the proper time of a moving body following that trajectory.

For infinitesimally separated events it is conventional to write

$$ds^2 = dx^2 + dy^2 + dz^2 - c^2 dt^2, \tag{2.16}$$

but beware of the notation: ds^2 might not be positive. Again, it is often more convenient to use an index notation and define $x^1 = x$, $x^2 = y$, and $x^3 = z$ in space, as well as $x^0 = ct$ for time. In this notation

$$ds^2 = -(dx^0)^2 + (dx^1)^2 + (dx^2)^2 + (dx^3)^2$$

$$= \begin{pmatrix} dx^0 & dx^1 & dx^2 & dx^3 \end{pmatrix} \begin{pmatrix} -1 & 0 & 0 & 0 \\ 0 & 1 & 0 & 0 \\ 0 & 0 & 1 & 0 \\ 0 & 0 & 0 & 1 \end{pmatrix} \begin{pmatrix} dx^0 \\ dx^1 \\ dx^2 \\ dx^3 \end{pmatrix},$$

---

[4] It is incorrect to conclude that Bob thinks that Alice's watch is running fast! We are measuring the time interval between two events at the same point in space in Bob's reference frame ($\Delta x' = 0$). They are not at the same point in space in Alice's reference frame: when $\Delta x' = 0$, $\Delta x = \gamma(v)v\Delta t' \neq 0$. The setup is not symmetric between Alice and Bob.

or

$$ds^2 = \eta_{\mu\nu} dx^\mu dx^\nu,$$

where $\mu, \nu = 0, 1, 2, 3$, and $\eta_{\mu\nu}$ are the components of the matrix[5]

$$\eta = \begin{pmatrix} -1 & 0 & 0 & 0 \\ 0 & 1 & 0 & 0 \\ 0 & 0 & 1 & 0 \\ 0 & 0 & 0 & 1 \end{pmatrix}. \tag{2.17}$$

The matrix η is the metric for Minkowski space-time.

Using the infinitesimal version of the Minkowski space line element (2.16), we can now calculate the length of *any* curve; just break it up into infinitely many infinitesimal segments and add them up to get an integral. This is most useful for the trajectories of physical objects following time-like curves for which $ds^2 < 0$ at all points along the curve (or, in extreme cases like photons, $ds^2 = 0$ along the curve). The length of any everywhere time-like curve Γ_{PQ} between any two points (events) P and Q with time-like separation is

$$S_{PQ} = \int_{\Gamma_{PQ}} \sqrt{(-ds^2)}. \tag{2.18}$$

The proper time

$$\tau_{PQ} = \frac{S_{PQ}}{c}$$

is then the time difference between the events as shown on a standard clock carried along the trajectory.

If we parameterise the curve by some parameter, λ say, and give the trajectory as functions $(x(\lambda), y(\lambda), z(\lambda),$ and $ct(\lambda))$, then the proper time τ along the trajectory is

$$\tau_{PQ} = \int_{\lambda_P}^{\lambda_Q} \sqrt{\left(\frac{dt}{d\lambda}\right)^2 - \frac{1}{c^2}\left\{\left(\frac{dx}{d\lambda}\right)^2 + \left(\frac{dy}{d\lambda}\right)^2 + \left(\frac{dz}{d\lambda}\right)^2\right\}} \, d\lambda.$$

---

[5] For the most part, when discussing a specific metric, we shall adhere to a convention in which Greek indices near the middle of the alphabet, $\mu, \nu, \ldots$, label coordinates in 4-dimensional space-time and take values $0, 1, 2, 3$, while Greek indices near the beginning of the alphabet $\alpha, \beta, \gamma, \ldots$ label coordinates in 3-dimensional space and take values $1, 2, 3$ (or sometimes just $1, 2$ in 2-dimensional examples). For more general discussions, where no particular dimension or metric is being considered, letters near the middle of the Greek alphabet will generally be used.

Indeed, we are free to choose λ to be the proper time along the trajectory, and we shall often make this choice, which imposes the condition

$$c^2 \left(\frac{dt}{d\tau}\right)^2 - \left\{\left(\frac{dx}{d\tau}\right)^2 + \left(\frac{dy}{d\tau}\right)^2 + \left(\frac{dz}{d\tau}\right)^2\right\} = c^2$$

on the four functions $t(\tau)$, $x(\tau)$, $y(\tau)$, and $z(\tau)$, equivalent to the statement that the 4-velocity has length $-c^2$, which is always true for a massive particle; the square of the 4-velocity must be negative (time-like 4-velocity) for any massive particle. The fact that the square of the 4-velocity is $-c^2$ does not mean that the 3-velocity is c; in fact, it must be strictly less than c for any massive object. If the 3-velocity in a given inertial reference frame is $\vec{u}$, then the 4-velocity has components $U^\mu = \gamma(u)(c^2, u^\alpha)$ and

$$\eta_{\mu\nu}U^\mu U^\nu = -c^2.$$

An important message to take away here is the following:

The magnitude of the length of an everywhere time-like curve in Minkowski space-time, when divided by the speed of light, equals the proper time of any body whose trajectory (world line) is that curve.

We shall use the same idea in curved space-times later. (This is an example of the use of the Equivalence Principle.)

A useful concept for visualising space-time is the notion of a *light cone*. Consider a flash of light leaving an event P with coordinates (ct_P, x_P, y_P, z_P). The flash will expand in a shell with radius $r = c(t - t_P)$ at time $t > t_P$. We can visualise this by suppressing the z coordinate and plotting the locus of all circles of radius $c(t - t_P)$ emanating from P. These trace out the surface of a cone radiating from P into the future, as shown in Figure 2.2. We can continue the cone into the past and imagine a shell of light converging on P. Such cones are called *light cones*, and each point (event) in space-time has its own light cone associated with it, as illustrated in Figure 2.3. Light cones are a useful way of visualising whether or not it is possible to communicate between two points in space-time. For a point Q to be able to receive a signal, travelling at the speed of light or less, from a point P, Q must lie somewhere inside the future light cone of P (and conversely P must lie somewhere inside the past light cone of Q). Only then can an event at P have any causal effect

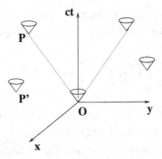

Figure 2.2 **Light cones in Minkowski space-time.** It is possible to send a signal from O to P; the straight line at 10 o'clock indicates the path of a light ray between these two events. It is not possible to send a signal from O to P' – that would require going outside of the light cone at O, which would require super-luminal speeds.

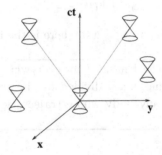

Figure 2.3 **Light cones in Minkowski space-time in both future and past directions.**

on the event at Q. The light cone structure of space-time is therefore sometimes called the *causal* structure. The structure of the light cones and the way the fit together in Minkowski space-time is fairly obvious, but it can become quite subtle in curved space-times.

2.5 The Hyperbolic Plane

Another interesting 2-dimensional geometry, which will be relevant later in our study of cosmology, is obtained by imposing a hyperbolic constraint,

$$x^2 + y^2 - z^2 = -a^2,$$

but not in Euclidean space, in 3-dimensional Minkowski space with line element

$$ds^2 = dx^2 + dy^2 - dz^2.$$

(We use z rather than ct here because there is no physical time – this is just a mathematical construction obtained by changing some signs.) Everything is symmetric under rotations about the z-axis, so it is perhaps better to use 2-dimensional polar coordinates with $\rho^2 = x^2 + y^2$ and write

$$\rho^2 - z^2 = -a^2,$$

$$ds^2 = d\rho^2 + \rho^2 d\phi^2 - dz^2. \tag{2.19}$$

Then,

$$2z dz = 2\rho d\rho \quad \Rightarrow \quad dz^2 = \frac{\rho^2 d\rho^2}{a^2 + \rho^2},$$

and putting this in (2.19) gives the 2-dimensional line element in (ρ, ϕ) coordinates,

$$ds^2 = \frac{a^2 d\rho^2}{a^2 + \rho^2} + \rho^2 d\phi^2 = a^2 \left(\frac{dr^2}{1 + r^2} + r^2 d\phi^2 \right),$$

where $r = \frac{\rho}{a}$. This is (2.13) with $K = -1$ and $0 \le r < \infty$. This space is called the Lobachevsky plane or the hyperbolic plane (or the Lobachevsky–Bolyai–Gauss geometry).[6] It was famously the first example of a consistent geometry that violates Euclid's fifth postulate (described in §4.4).

To visualise the geometry of the Lobachevsky plane and compare it to the sphere and the flat plane, we make yet another coordinate transformation and define u by

$$r = \sinh u.$$

Then,

$$ds^2 = a^2 (du^2 + \sinh^2 u \, d\phi^2), \tag{2.20}$$

which is very like (2.9), but with $\sin \theta$ replaced by $\sinh u$ in the second term. In (2.9), circles of constant latitude, that is, constant θ, have radius

---

[6] The hyperbolic plane was discovered at pretty much the same time, by Janós Bolyai in Hungary and Nikolai Lobachevsky in Russia. Nikolai Ivanovich Lobachevsky was further immortalised in a Tom Lehrer song of the same name (though Lehrer's accusation of plagiarism is invented – he just used Lobachevsky's name because he thought it went well with the metre of the tune).

Figure 2.4 **The geometry of a sphere.**

$a\sin\theta \leq a\theta$ and grow in size more slowly than linearly in θ, while in
(2.12) (or (2.13), with $K = 0$) circles of constant $\tilde{r}$ have radius $\tilde{r}$ and
grow linearly with $\tilde{r}$.

This means, as any map maker knows, that trying to represent the
northern hemisphere of a sphere as a disc on a flat plane, with the north
pole at the centre of the disc, distorts objects near the equator and
stretches them. The curvature of a sphere can be detected by the way
that it reflects light, as shown in Figure 2.4.

With the line element in (2.20), circles of constant u have radius
$a\sinh u$ and thus grow faster than linearly in u; indeed, they grow
exponentially for large u. As a result, any attempt to represent the Lobachevsky plane on a flat plane distorts objects at large u and *shrinks* them
(see Figure 2.5).

To summarise, we can express three different geometries in one
formula, (2.13), with

$$K = +1 \quad \Leftrightarrow \quad ds^2 = a^2(d\theta^2 + \sin^2\theta\, d\phi^2) \qquad \text{sphere}$$
$$K = 0 \quad \Leftrightarrow \quad ds^2 = a^2(dr^2 + r^2 d\phi^2) \qquad \text{flat plane}$$
$$K = -1 \quad \Leftrightarrow \quad ds^2 = a^2(du^2 + \sinh^2 u\, d\phi^2) \qquad \text{hyperbolic plane.}$$

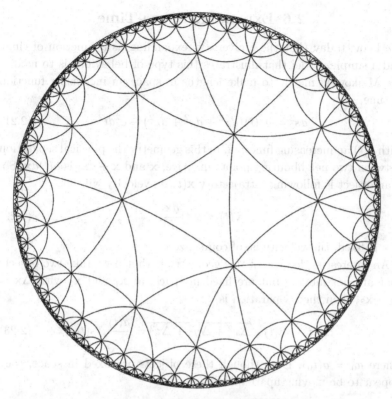

Figure 2.5 **A representation of the hyperbolic plane, a space with constant negative curvature.** Objects of equal size look smaller nearer the edge, which is the opposite of what happens for the sphere in Figure 2.4 (which has positive curvature). In this geometry the edges of the triangles are geodesics.

Higher-dimensional versions of these will be important when we discuss the geometry of the Universe on cosmological length scales in Chapter 7.

All the examples so far have been either flat or obtained by studying surfaces embedded in a higher-dimensional flat space by imposing algebraic conditions like (2.7). However, it is important to note that it is not actually necessary to have an embedding into a higher-dimensional space in order to have curvature; curved spaces can be completely characterised just by writing down a line element, without reference to higher dimensions.

2.6 Expanding Space-Time

We know today that the Universe is expanding as a function of time, and a simple model that illustrates this type of behaviour is to modify the Minkowski metric to make lengths in space an increasing function of time,

$$ds^2 = a^2(t)\big(dx^2 + dy^2 + dz^2\big) - c^2dt^2, \qquad (2.21)$$

with $a(t)$ an increasing function. In this geometry the physical separation between two neighbouring points in space, $\mathbf{x}$ and $\mathbf{x} + d\mathbf{x}$, is $a(t)d\mathbf{x}$. So, if an object is following a trajectory $\mathbf{x}(t)$ its velocity will be

$$\mathbf{v}(t) = a(t)\frac{d\mathbf{x}}{dt}. \qquad (2.22)$$

If $\mathbf{x}$ is fixed, the velocity is, of course, zero.

An interesting feature of this geometry is that if we take two objects that are not moving but are fixed in space, at $\mathbf{x}_1$ and $\mathbf{x}_2$ with $\Delta\mathbf{x} = \mathbf{x}_1 - \mathbf{x}_2$, then their separation is

$$\Delta s(t) = a(t)\sqrt{\Delta x^2 + \Delta y^2 + \Delta z^2} = \frac{a(t)}{a_0}\Delta s(t_0), \qquad (2.23)$$

where $a_0 = a(t_0)$. Even though these objects are fixed in space, they appear to be moving apart with speed

$$\frac{d\Delta s}{dt} = \frac{\Delta s(t_0)}{a_0}\frac{da}{dt}. \qquad (2.24)$$

Note that for any non-constant $a(t)$, as long as $\frac{da}{dt} > 0$, we can always choose $\Delta s(t_0)$ large enough (we could be discussing two galaxies, for example) to make $\frac{d\Delta s}{dt} > c$, which shows that this cannot be a physical speed. Δs is calculated at a constant time $\Delta t = 0$, and no physical object can follow a trajectory with $\Delta t = 0$ (time stops for no man, or galaxy), so this is not the physical velocity of a galaxy: it is an apparent velocity.

Nevertheless, there are physical consequences: suppose $\mathbf{x}_1$ and $\mathbf{x}_2$ represent the positions of two galaxies at rest in this geometry and $\frac{da}{dt} > 0$. Then $\frac{d\Delta s(t)}{dt} > 0$ and the galaxies will appear to be moving apart. This is the case in our universe, and this is the reason we say that the Universe is expanding. Note that there is no expansion away from any particular point; space is perfectly homogeneous and all points are equivalent as far as the geometry is concerned – there is no preferred origin.

Although $\frac{d\Delta s(t)}{dt}$ is not a physical velocity, we can still use it to derive an important physical consequence. If $\frac{da}{dt} > 0$ and $\Delta s(t_0)$ is large enough, then $\frac{d\Delta s(t)}{dt} > c$ and it will not be possible for light to travel from galaxy 2 to galaxy 1. Galaxy 2 will be hidden from the sight of galaxy 1, and vice versa, though it may become visible in the future, if $\frac{da}{dt}$ becomes smaller in the future, and it may have been visible in the past if $\frac{da}{dt}$ has been smaller in the past. (The question of whether particular galaxies were visible in the past, or will be visible in the future, depends on the functional form of $a(t)$.)

It is emphasised that the space coordinates and their differences in (2.23) are fixed; the galaxies are not moving. This gives us license to make Zen-like statements which are difficult to resist, such as 'galaxies that are not moving fly apart from one another'. The geometry of curved space-times can indeed be very strange.

2.6.1 The Hubble Expansion

Consider a beam of light passing between two galaxies, A and B. Assuming space is homogeneous and isotropic (i.e. it looks the same from every point and in every direction), we are free to choose an origin anywhere in 3-dimensional space and we can use spherical polar coordinates to describe the positions of the galaxies. Choose the origin to be at galaxy B and orient the axes so that galaxy A is on the x-axis, fixed at $r = r_A$ with $\phi_A = 0$ and $\theta_A = \frac{\pi}{2}$. The physical distance between the galaxies, at fixed t, is $a(t)r_A$, where $a(t)$ is the cosmological scale factor.

A beam of light travelling radially inwards from A to B follows a light-like trajectory with $d\theta = d\phi = 0$ satisfying

$$0 = -c^2 dt^2 + a(t)^2 dr^2.$$

At any point along the light beam's trajectory, the light travels a distance $-a(t)\delta r$ in the time $\delta t = -\frac{a(t)\delta r}{c}$. Suppose light leaves A at time t_1 and arrives at B at time t_2 (B could be our Galaxy and $t_2 = t_0$ the present day); then we can integrate radially inwards along the light beam's trajectory to get

$$\int_{t_1}^{t_2} \frac{dt}{a(t)} = -\frac{1}{c} \int_{r_A}^{0} dr = \frac{r_A}{c}. \tag{2.25}$$

At a later time, light leaving A at $t_1 + \Delta t_1$ would reach B at $t_2 + \Delta t_2$,

$$\int_{t_1+\Delta t_1}^{t_2+\Delta t_2} \frac{dt}{a(t)} = -\frac{1}{c} \int_{r_A}^{0} dr = \int_{t_1}^{t_2} \frac{dt}{a(t)},$$

so

$$\int_{t_1+\Delta t_1}^{t_2+\Delta t_2} \frac{dt}{a(t)} = \int_{t_1}^{t_2} \frac{dt}{a(t)} \quad \Rightarrow \quad \int_{t_2}^{t_2+\Delta t_2} \frac{dt}{a(t)} = \int_{t_1}^{t_1+\Delta t_1} \frac{dt}{a(t)}.$$

Take Δt_1 and Δt_2 to be the inverses of optical frequencies, ν_1 and ν_2 say, and assume $a(t) = a(t_1)$ is essentially constant between t_1 and $t_1 + \Delta t_1$ and $a(t) = a(t_2)$ between t_2 and $t_2 + \Delta t_2$. (This always is an extremely good approximation: for optical frequencies, $\Delta t \approx 10^{-15}$s, while the scale factor $a(t)$ only changes appreciably on time scales of order 10^9 yrs $\approx 3 \times 10^{16}$s.) Then

$$\frac{1}{a(t_1)}\Delta t_1 = \frac{1}{a(t_2)}\Delta t_2$$

$$\frac{a(t_1)}{a(t_2)} = \frac{\Delta t_1}{\Delta t_2} = \frac{\nu_2}{\nu_1}. \tag{2.26}$$

If $a(t_2) > a(t_1)$, then $\nu_2 < \nu_1$ (i.e. the light is red-shifted).

If we furthermore make the stronger assumption[7] that $a(t)$ is slowly varying on the timescale $t_2 - t_1$ and is an analytic function of time, then it can be Taylor expanded around t_2 as

$$a(t_1) = a(t_2)\{1 - (t_2 - t_1)H(t_2) + \ldots\}, \quad \text{where} \quad H(t_2) = \frac{\dot{a}}{a}\bigg|_{t_2}$$

$$\Rightarrow \quad a(t_2) = a(t_1)\{1 + (t_2 - t_1)H(t_2) + \ldots\},$$

where the dots represent a negligible contribution, provided $(t_2 - t_1)H(t_2) \ll 1$. If t_2 is the present day, t_0, then $H_0 := \frac{\dot{a}}{a}\big|_{t_0}$ is called the *Hubble constant*, although $\frac{\dot{a}}{a}$ itself is not a constant. Suppose light left a star in a distant galaxy far, far away at a time $t = t_1$ and arrives at our telescope at the present day $t_0 = t_2$. Let $s_{AB} = c(t_2 - t_1)$. (We could call this the time-of-flight distance, the distance as determined by the time it takes the light to travel from A to B divided by the speed of light.) Then, assuming $t_0 - t \ll H_0^{-1}$, we have

$$z := \frac{\delta\nu}{\nu_2} = \frac{\nu_1 - \nu_2}{\nu_2} = \frac{\nu_1}{\nu_2} - 1 = \frac{a(t_2)}{a(t_1)} - 1 = (t_0 - t)H_0 = \frac{s_{AB}H_0}{c}. \tag{2.27}$$

z is called the *red-shift* of the galaxy. It is a measure of the amount by which blue or yellow light emitted from a distant galaxy is shifted

---

[7] For example, $t_2 - t_1$ would be 10^7 years for light passing between two galaxies 10 million light years apart and, as already mentioned, the scale factor $a(t)$ only changes appreciably on time scales of order 10^9 yrs.

towards the red end of the spectrum (i.e. towards longer wavelengths) by the time it reaches us. z is easily measured by identifying spectral lines of a specific type of atom, usually hydrogen, in a galaxy's spectrum, calculating its apparent frequency (ν_2) and comparing this frequency to that of an atom at rest in the laboratory, which would be the same as ν_1.

We have

$$z = \frac{H_0}{c} s_{AB}, \tag{2.28}$$

giving the *red-shift-distance relation*

$$\boxed{s_{AB} = \frac{c}{H_0} z.} \tag{2.29}$$

H_0 is thus the constant of proportionality in a linear relation between the red-shift z and the distance s_{AB}, hence the name Hubble constant. But we emphasise that $H(t) = \frac{\dot{a}}{a}$ is not a constant in time; the Hubble constant is $\frac{\dot{a}}{a}$ at the present day $H_0 = H(t_0) = \frac{\dot{a}}{a}\big|_{t_0}$.

Equation (2.29) is also commonly known as the *Hubble relation* after Edwin Hubble who published it in 1929, though much of the work that led to Hubble's discovery was done by his assistant Milton L. Humason.[8]

H_0 has dimensions of (time)$^{-1}$ but is usually quoted as km s^{-1}Mpc^{-1}. (A parsec is an astronomical unit of distance that is 3.26 light years; a mega parsec (Mpc) is therefore about 3 million light years.) It can be calculated by measuring the red-shifts of many galaxies and estimating their distance, from the brightness of standard objects in the galaxies, such as variable stars called Cepheid variables,[9] or from certain kinds of supernovae (technically type Ia supernovae) whose absolute brightness can be inferred from the rate at which they fade after they explode. The Hubble constant can be measured by plotting the distance to the galaxy against its red-shift, as shown in Figure 2.6; it is essentially the slope of this line, though deviations from linearity are visible around $z = 0.5$

---

[8] Humason started his scientific career as a mule driver ferrying materials up to the Mount Wilson Observatory.

[9] These are stars whose brightness varies in a regular periodic way with a period which is correlated with their average brightness. By measuring their period we know their intrinsic luminosity and can estimate the distance to the galaxy from their apparent luminosity – the farther away the galaxy, the fainter the star. Cepheid variables were discovered by the American astronomer Henrietta Swan Leavitt in 1908 and were named after the first of their type to be identified, in the constellation of Cepheus. If she had been a man, they would probably have been called Leavitt variables.

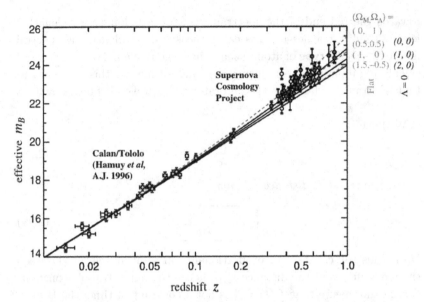

Figure 2.6 **Hubble diagram.** The horizontal axis is the red-shift, the vertical axis is a measure of the apparent luminosity, which is related to d_A. H_0 can be estimated from the slope of the line. Deviations from linearity become apparent when z approaches 0.5 (the significance of Ω_M and Ω_Λ is explained in Chapter 7).
Adapted from Perlmutter *et al.*, *Astrophys. J.* **517** (1999) 565, © AAS. Reproduced with permission.

and greater. The current best estimate of H_0 from observations like this is[10]

$$H_0 = 73.2 \pm 1.3 \, \mathrm{km\,s^{-1}Mpc^{-1}} \tag{2.30}$$
$$:= h \times 100 \, \mathrm{km\,s^{-1}Mpc^{-1}} \quad \text{where} \quad h = 0.732 \pm 0.013.$$

This means that a galaxy at a distance of 1 Mpc exhibits a red-shift corresponding to a velocity of about 73 $\mathrm{km\,s^{-1}}$. The linear relation (2.29) is only valid for 'small' $(t_2 - t_1)$, that is, red-shifts $z \ll 1$.

The preceding analysis has assumed that the two galaxies are not moving, in the sense that their spatial coordinates are fixed, and this is never the case for real galaxies. Galaxies do tend to move around in

---

[10] There is more uncertainty in H_0 than is apparent from the errors quoted here. There are different ways of determining H_0 and another method, which will be discussed in more detail in §7.1, gives a slightly lower value of $67.7 \pm 0.4 \mathrm{km\,s^{-1}Mpc^{-1}}$.

space in a somewhat random way, under the influence of the gravity of other nearby galaxies, and we can try to take account of this by studying the average motion of clusters of galaxies that are gravitationally bound to each other. For example, there is a cluster of galaxies in the constellation of Virgo ($\approx 2,500$ galaxies) with an average red-shift $c\frac{\delta\nu}{\nu} = 1,150\,\mathrm{kms}^{-1} \Rightarrow \frac{\delta\nu}{\nu} = 0.00383$, giving $s_{AB} = \frac{c}{H_0}\frac{\delta\nu}{\nu} = 17\,\mathrm{Mpc} = 56\,\mathrm{Mlyrs}$. The current red-shift record is $\frac{\delta\nu}{\nu} \approx 10$, which gives a naïve red-shift-distance relation of $s_{AB} \approx 4.5 \times 10^4\,\mathrm{Mpc}$, but this is not the true physical distance because $\frac{\delta\nu}{\nu} > 1$ is not small. For z near or above 1, we need to know the functional form of $a(t)$ to determine the distance from the red-shift.

$\frac{1}{H_0}$ has dimensions of time: $\frac{1}{H_0} \approx 4 \times 10^{17}\mathrm{s} \approx 13$ billion years which is the approximate age of the Universe (more accurate modelling gives $t_0 = 13.8 \times 10^9$ years). As an order of magnitude estimate, $a_0 = \frac{1}{H_0}c \approx 10^{26}\mathrm{m}$ is the approximate size of the observable Universe.

2.6.2 The Milne Universe

Even flat space-time can appear to be expanding when viewed from a certain perspective. We obtained the Minkowski metric in §2.4 by starting with the Euclidean metric and changing a sign; in two dimensions, for example,

$$ds^2 = dx^2 + dy^2 \quad \longrightarrow \quad -dx^2 + dy^2 \quad \underset{x\,\to\,ct}{\longrightarrow} \quad -c^2dt^2 + dy^2$$

gives the line element on 2-dimensional Minkowski space-time with space coordinate y (one space dimension). If, however, we use 2-dimensional polar coordinates and send r to ct, like so

$$ds^2 = dr^2 + r^2d\phi^2 \quad \longrightarrow \quad -dr^2 + r^2d\phi^2 \quad \underset{r\,\to\,ct}{\longrightarrow} \quad -c^2dt^2 + c^2t^2d\phi^2,$$

then we get a 2-dimensional space-time (one space dimension) in which space is expanding linearly with time, $a(t) = ct$, a simple model for expanding space-time called the *Milne universe*.

To visualise the Milne universe, 'unwrap' the polar coordinate ϕ by choosing a fixed time t_0 and letting

$$y = ct_0\phi$$

range from $-\infty < y < \infty$. Then

$$ds^2 = -c^2dt^2 + \left(\frac{t}{t_0}\right)^2 dy^2.$$

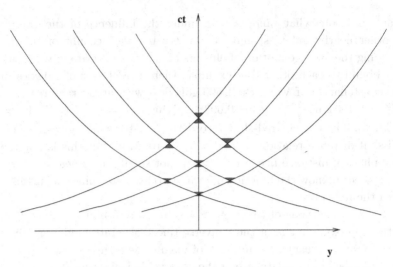

Figure 2.7 **The Milne universe.** Light cone (causal) structure of the
Milne universe.

A beam of light leaving y_0 at time t_0 follows a trajectory with $ds^2 = 0$
or

$$\frac{dt}{dy} = \pm\frac{t}{ct_0} \quad \Rightarrow \quad t = \exp\left\{\pm\left(\frac{y-y_0}{ct_0}\right)\right\}.$$

Plotting exponential curves $t(y)$ for various values of y_0 and t_0 gives the
light cone structure of the Milne universe, as shown in Figure 2.7.

2.7 Gravitational Red-Shift and the Global Positioning System

Our next metric is that of a space-time outside a static compact spherically symmetric mass, such as a non-rotating planet or star. Such an
object with mass M would give rise to a spherically symmetric gravitational field in Newtonian gravity – according to Einstein it merely
curves space-time. The metric can be derived from Einstein's field equations for general relativity, and these will be discussed later; for the
moment we just write it down. Using spherical polar coordinates (r, θ, ϕ)
in 3-dimensional space, with the mass M at the origin $r = 0$, the line
element is

$$ds^2 = -\left(1 - \frac{2GM}{c^2r}\right)c^2dt^2 + \frac{dr^2}{\left(1 - \frac{2GM}{c^2r}\right)} + r^2\left(d\theta^2 + \sin^2\theta\,d\phi^2\right). \quad (2.31)$$

Here G is Newton's universal constant of gravitation; the factors of G and c^2 are necessary for dimensional reasons to render $\frac{GM}{c^2 r}$ dimensionless. Note that this reduces to the flat space-time Minkowski metric (3.15) when $M = 0$; as one would expect, there is no gravitational field, and space-time is flat when there is no mass. The metric associated with the line element (2.31) is known as the *Schwarzschild metric*, after Karl Schwarzschild, who first derived it in 1915, the same year that Einstein produced his general theory of relativity.[11] This geometry mimics that of the gravitational field outside a spherically symmetric mass M centred on $r = 0$.

Note that the coefficients of both dt^2 and dr^2 in (2.31) change sign as r passes through the value $r_S = \frac{2GM}{c^2}$. Indeed, the coefficient of dt^2 vanishes and the coefficient of dr^2 diverges at $r = r_S$; something very funny is going on there. r_S is a characteristic length scale associated with this geometry called the *Schwarzschild radius*. Conversely, the coefficient of dt^2 diverges and the coefficient of dr^2 vanishes at $r = 0$. We shall see later that the pathologies at $r = r_S$ are due to the fact that we have chosen a bad set of coordinates here; the singularity is just a coordinate singularity, and the geometry is perfectly regular there. The singularity at $r = 0$ is, however, another matter; the geometry is genuinely pathological there and tidal forces become infinite. For the moment we shall avoid these issues by simply restricting $r > r_S$.

Gravitational Red-Shift

The Schwarzschild line element (2.31) describes space-time outside of a spherically symmetric non-rotating star or planet. We can ask what this space-time looks like to an observer a long way away, $r \to \infty$. Suppose we have an observer fixed at $r \to \infty$ and consider a time interval Δt; then her proper time is

$$\Delta \tau_\infty = \lim_{r \to \infty} \left(1 - \frac{2GM}{c^2 r}\right)^{\frac{1}{2}} \Delta t = \Delta t.$$

Now suppose a man with a large expensive watch is standing directly below her at a fixed point r_1, with $r_1 > r_S$ but finite. The woman cannot keep her eyes off his fancy watch. We can set $dr = d\theta = d\phi = 0$ for the man and interpret

[11] Schwarzschild derived this metric while serving in the German army during World War I – sadly he died a year later in 1916 from an illness contracted at the Russian front.

$$\Delta\tau_1 = \frac{\sqrt{|ds^2|}}{c} = \left(1 - \frac{2GM}{c^2 r_1}\right)^{\frac{1}{2}} \Delta t = \left(1 - \frac{2GM}{c^2 r_1}\right)^{\frac{1}{2}} \Delta\tau_\infty$$

as the proper time on his watch at r_1 associated with the time interval Δt at infinity. Since

$$\Delta\tau_1 = \left(1 - \frac{2GM}{c^2 r_1}\right)^{\frac{1}{2}} \Delta\tau_\infty < \tau_\infty,$$

the clock at infinity is ticking faster than the watch at r_1 – time slows down in a gravitational field. This is known as *gravitational red-shift* or *gravitational time dilation*.

In fact, as $r_1 \to r_S$, $\Delta\tau_\infty \to \infty$ for any $\Delta\tau_1$, no matter how small: time is frozen, and the woman at infinity waits an eternity for the watch to tick once. She thinks his flashy watch has stopped and is singularly unimpressed by it. A corollary of this is that an observer a long way from r_S can never see any event that occurs at r_S (or at $r < r_S$). For this reason the spherical surface at $r = r_S$ is called an *event horizon*; we shall see later that no event that occurs at $r < r_S$ can ever be seen by an observer at $r > r_S$.

Now suppose our female observer is at a fixed but finite r_E and the watch is fixed at $r_1 > r_E$. (For example, r_E might be the surface of the Earth and r_1 at the top of a hill 1000 m high, such as Schiehallion in Scotland where G was first measured.) Then the woman's proper time τ_E is now related to the proper time on the face of the watch τ_1 by

$$\frac{\Delta\tau_E}{\sqrt{1 - \frac{2GM}{c^2 r_E}}} = \frac{\Delta\tau_1}{\sqrt{1 - \frac{2GM}{c^2 r_1}}}. \tag{2.32}$$

The mass of the Earth is $M = 5.972 \times 10^{24}$ kg and its radius is $r_E = 6{,}371$ km, so its Schwarzschild radius is just short of a centimetre, $r_S = 8.870$ mm. $\frac{r_S}{r_E} = 1.392 \times 10^{-9}$ is tiny and, if $r_1 > r_E$, $\frac{r_S}{r_1}$ is even smaller. We can therefore approximate

$$\Delta\tau_E = \sqrt{\frac{1 - \frac{r_S}{r_E}}{1 - \frac{r_S}{r_1}}} \Delta\tau_1 \approx \left(1 - \frac{r_S}{2}\left(\frac{1}{r_E} - \frac{1}{r_1}\right)\right) \Delta\tau_1,$$

with an error of less than one part in a billion. If the watch watch is at a height with $h \ll r_E$, then $r_1 = r_E + h$ and we can also use the approximation

$$\frac{1}{r_E} - \frac{1}{r_1} = \frac{1}{r_E} - \frac{1}{r_E + h} = \frac{1}{r_E} - \frac{1}{r_E\left(1 + \frac{h}{r_E}\right)} \approx \frac{h}{r_E^2}.$$

Hence

$$\Delta\tau_E \approx \left(1 - \frac{1}{2}\frac{r_S h}{r_E^2}\right)\Delta\tau_1.$$

The time on the female observer's watch is ever so slightly less than on the watch fixed at h – the watch ticks slightly faster and again time slows down in a gravitational field. For example, if $h = 1,000$ m, then $\frac{h}{r_E} = \frac{1}{6371} = 1.57 \times 10^{-4}$, and $\frac{r_S}{r_E} = 1.392 \times 10^{-9}$ gives

$$\Delta\tau_E = \left(1 - (1.09 \times 10^{-13})\right)\Delta\tau_1.$$

This may not seem much, but if a watch is left on the top of the hill for a year $(3 \times 10^7\,\text{s})$ it will gain $3.3\,\mu\text{s}$ relative to a clock on the ground, which is measurable.

This experiment has in fact been done, though with the watch at the top of a building 22.5 m tall and the observer in the basement. It is not even necessary to wait a year if you don't really use a simple watch. The time dilation manifests itself as a frequency shift in light – photons emitted from an atom (light) or a radioactive atomic nucleus (γ-rays) have a characteristic frequency, and that frequency decreases when time is dilated (they lose energy as they climb up out of the gravitational field); conversely, the frequency increases as photons fall into a gravitational field. A γ-ray photon emitted downwards from a source at the top of a building has a slightly higher frequency when it reaches the basement, by 1 part in 2.5×10^{-15} for a building 22.5 m tall; but even this is measurable using a very sensitive technique called Mössbauer spectroscopy, and this was done by Pound and Rebka at Harvard in 1959. A more sensitive version of the experiment, named gravity probe A, was performed in 1976, involving a hydrogen maser launched vertically to a height of 10,226 km, increasing the sensitivity to 1 part in 4.3×10^{-10}.

The gravitational time-dilation effect featured in the 2014 film 'Interstellar', in which astronauts land on the surface of a planet (Miller's planet) orbiting close to a supermassive black hole (Gargantua) 100 million times the mass of the Sun and 10 billion light years from Earth. One hour of their time corresponded to seven years on Earth – a time dilation factor of 61,000.

The Global Positioning System
Gravitational time dilation must be accounted for in the Global Positioning System (GPS). Given the speed of light, if we want positional

accuracy to 10 m, we require timing to 30 ns, and gravitational red-shift effects are greater than this.

GPS satellites are typically in orbits that circle the Earth twice a day. If the orbit is circular, then

$$\frac{GM}{r^2} = \omega^2 r \qquad \Rightarrow \qquad \omega^2 = \frac{GM}{r^3},$$

where ω is the angular frequency, so the period is

$$T = \frac{2\pi}{\omega} = 2\pi\sqrt{\frac{r^3}{GM}}.$$

This is just Kepler's third law – the square of the period is proportional to the cube of the radius of the orbit. The mass of the Earth is $M = 5.972 \times 10^{24}$ kg, so, if we want a 12-hour orbit, 12 hours $= 4.32 \times 10^4$ s, the radius must be $r = 26{,}600$ km.

In a time Δt the satellite moves through an angle $\Delta\phi = \omega\Delta t = \frac{v}{r}\Delta t$ with $v = \frac{2\pi r}{T} = 3.9 \times 10^3 \text{ms}^{-1}$, just under 4 km per second. Fix $r = 26{,}600$ km and choose the plane of the satellite's orbit to be in the fixed $\theta = \frac{\pi}{2}$ plane; then the Schwarzschild line element (2.31) is

$$\Delta s^2 = -c^2\left(1 - \frac{2GM}{c^2 r}\right)\Delta t^2 + r^2\Delta\phi^2 = -c^2\left(1 - \frac{2GM}{c^2 r} - \frac{v^2}{c^2}\right)\Delta t^2.$$

So, the proper time on the satellite's clock is

$$\Delta\tau_S = \sqrt{\left(1 - \frac{2GM}{c^2 r} - \frac{v^2}{c^2}\right)}\,\Delta t.$$

There are two distinct effects here: with $M = 0$, $\sqrt{1 - \frac{v^2}{c^2}}$ is special relativistic time-dilation; with $v = 0$, $\sqrt{1 - \frac{2GM}{c^2 r}}$ is gravitational red-shift.

In the same Δt the proper time on a clock for an observer fixed on the surface of the Earth is

$$\Delta\tau_E = \sqrt{\left(1 - \frac{2GM}{c^2 r_E}\right)}\,\Delta t.$$

(We ignore the rotation of the Earth; the relevant $\frac{v^2}{c^2}$ term for that is negligible.) The clock on the satellite and the clock on the Earth are ticking at different rates:

$$\Delta\tau_S = \frac{\sqrt{1 - \frac{2GM}{c^2 r} - \frac{v^2}{c^2}}}{\sqrt{1 - \frac{2GM}{c^2 r_E}}}\,\Delta\tau_E \approx \left(1 - \frac{r_S}{2r} + \frac{r_S}{2r_E} - \frac{1}{2}\frac{v^2}{c^2}\right)\Delta\tau_E, \quad (2.33)$$

and this must be taken into account when performing trigonometric calculations to determine the observer's position on the surface of the Earth using the satellite's signal. Now $\frac{v}{c} = 1.3 \times 10^{-5}$, so $\frac{1}{2}\frac{v^2}{c^2} = 0.85 \times 10^{-10}$ and, using special relativistic time dilation, this translates to a decrease of $7.3\,\mu s$/day in (2.33). Multiplying by the speed of light, this converts to a drift of $2.2\,km$/day.

On the other hand, with $r = 26{,}600\,km$, $r_E = 6{,}380$ km, and $r_S = 8.9$ mm,

$$\frac{r_S}{2r_E} - \frac{r_S}{2r} = \frac{r_S}{2}\left(\frac{r - r_E}{rr_E}\right) = 5.3 \times 10^{-10}.$$

Using gravitational red-shift, this translates to an increase of $45.8\,\mu s$/day in (2.33), or $13.7\,km$/day. These two effects act in opposite directions, but the gravitational red-shift wins.

Even before the GPS was developed, this kind of red-shift had been detected in an experiment in 1971 by Hafele and Keating, who put an atomic clock into a commercial airliner flying around the world and, when it returned, compared its time to that of an identical clock which was sitting stationary in their laboratory. After the flight, the flying clock was 270 ns ahead of the stationary clock if the plane was flying westwards, and 60 ns behind it if the plane was flying eastwards. The results differ due to the special relativistic effect arising from the rotation of the Earth, which is more significant in this case.

2.8 Tides – the Relativistic Map Maker's Problem

There is a relativistic version of the map maker's problem described on page 28. The map maker can choose coordinates on a curved surface so that it looks flat over a small enough region, and there are analogues of this in curved spaces in higher dimensions. In 4-dimensional space-time an observer can choose coordinates at a point on his world line (his birth, for example) so that space-time looks flat locally, like Minkowski space-time, as long as he does not look at events too far away from his chosen event. This is known as a *locally* inertial coordinate system. It can be considered to be an inertial coordinate system for events close enough to the chosen point. However, if the observer looks at events further away and there is a non-uniform gravitational field (curvature), then he will see tides.

But he can do even better than this. If he is in free fall he can find coordinates so that space-time looks locally flat on his whole world line.

However, no man is a point (with apologies to John Donne). Suppose we have a black hole a hundred times more massive than the Earth, with a Schwarzschild radius of 1 m. An astronaut in free fall does not feel any acceleration and does not weigh anything, so he will not be crushed by his own weight. But he's still in trouble – as he approaches the black hole there will be huge tides, over distances of a metre. If he falls feet first and is 2 metres tall, his feet will be pulled more strongly than his head – he will be stretched into spaghetti, ripped apart by tidal forces before he even reaches the event horizon. The relativistic map maker's problem can be lethal.

Problems

1) In the y-z plane a curve is given by the equation

$$y = b \left(1 + \cos(z/a)\right) \qquad x = 0 \qquad -a\pi \leq z \leq a\pi,$$

where a and b are positive constants. Sketch the surface obtained by rotating this curve around the z-axis, and derive the metric

$$ds^2 = \left(\frac{a^2 + 2b\rho - \rho^2}{\rho(2b - \rho)}\right) d\rho^2 + \rho^2 d\phi^2.$$

Do you think the singularities at $\rho = 0$ are just coordinate singularities or a sign of something more severe?

2) Show that two space-like separated events in Minkowski space-time can have their order in time reversed by a Lorentz transformation; that is, if P and Q are two events with space-like separation, and P happens before Q in one inertial reference frame, then there exists an inertial reference frame in which Q happens before P.

3) Express the 2-dimensional Minkowski metric

$$ds^2 = dx^2 - c^2 dt^2$$

in terms of the coordinates

$$v = ct + x$$
$$u = ct - x$$

(called light cone coordinates).

Sketch the lines of constant u and v in the ct-x plane.

4) Express the 2-dimensional Minkowski metric

$$ds^2 = dx^2 - c^2 dt^2$$

in terms of new coordinates T and X defined by

$$ct = X \sinh\left(\frac{cT}{L}\right)$$

$$x = X \cosh\left(\frac{cT}{L}\right),$$

where L is a fixed length.

Sketch the lines of constant T and X in the ct-x plane. The coordinates (cT, X) are called *Rindler* coordinates, and curves of constant X correspond to the world lines of uniformly accelerating observers in Minkowski space-time.

5) When 2-dimensional polar coordinates $(x, y) = (\rho \cos\phi, \rho \sin\phi)$ are used for the space-like part, the metric for 3-dimensional Minkowski space-time is

$$ds^2 = d\rho^2 + \rho^2 d\phi^2 - c^2 dt^2,$$

where $\rho^2 = x^2 + y^2$.

Show that the metric on the hyperboloid $\rho^2 - c^2 t^2 = L^2$, where L is a fixed length, is given by

$$ds^2 = (L^2 + c^2 t^2)d\phi^2 - \frac{L^2 c^2 dt^2}{L^2 + c^2 t^2}.$$

Sketch the surface in (x, y, ct)-space.

Re-express the line element on the hyperboloid in terms of (cT, ϕ) with

$$ct = L \sinh\left(\frac{cT}{L}\right)$$

and show that it represents an expanding 2-dimensional universe,

$$ds^2 = a^2(T)d\phi^2 - c^2 dT^2,$$

and determine the function $a(T)$. This is called de Sitter space-time. A 4-dimensional version of this was one of the first models for an expanding space-time that was proposed.

6) One can use any coordinate system one likes to describe physics; the physics should be independent of the coordinate system. But the choice of coordinates can strongly influence the way we think of things.

A wildlife photographer leaves her tent in the morning and walks 10 km north, where she encounters bird tracks heading east. She follows the tracks for 1 km, keeping an easterly direction, and catches up with the bird to photograph it. She then travels 10 km south and arrives back at her tent.

What species was the bird?

3

Geodesics

In any given geometry the shortest path between two points is called a geodesic. Geodesics are of central importance in general relativity, as they correspond to the trajectories of freely falling objects, including, of course, freely falling observers. In this chapter we shall learn how to determine the geodesics associated with a given metric and look at a number of instructive examples.

3.1 Euclidean Geometry

The simplest geometry is 2-dimensional Euclidean space. Let $(x(\tau), y(\tau))$ denote coordinates on a path from initial point $\mathbf{x}_0 = (x_0^1, x_0^2)$ at $\tau = \tau_0$ to final point $\mathbf{x}_1 = (x_1^1, x_1^2)$ at $\tau = \tau_1$, parameterised by τ with $\mathbf{x}_0 = \mathbf{x}(\tau_0)$ and $\mathbf{x}_1 = \mathbf{x}(\tau_1)$. The infinitesimal line element at the point labelled by τ is

$$ds = \sqrt{(dx^1)^2 + (dx^2)^2} = \left(\sqrt{\left(\frac{dx^1}{d\tau}\right)^2 + \left(\frac{dx^2}{d\tau}\right)^2} \right) d\tau.$$

The total length of the path is

$$L_{01} = \int_{\mathbf{x}_0}^{\mathbf{x}_1} ds = \int_{\tau_0}^{\tau_1} \left(\sqrt{\left(\frac{dx^1}{d\tau}\right)^2 + \left(\frac{dx^2}{d\tau}\right)^2} \right) d\tau.$$

We can find the shortest path between the two fixed points $\mathbf{x}_0$ and $\mathbf{x}_1$ by varying $\mathbf{x}(\tau)$, keeping the end points fixed, and looking for extrema of L_{01}. The square root makes things a bit messy, and it is easier instead to look for extrema of[1]

---

[1] Finding the extrema of a function $f(x)$ is equivalent to finding the extrema of $f^2(x)$ as long as $f(x)$ never changes sign, which is necessarily the case if $f(x)$ is never zero.

44

$$S[x^\alpha; \dot{x}^\alpha] = \int_{\tau_0}^{\tau_1} \mathcal{L}(x^\alpha; \dot{x}^\alpha) d\tau, \qquad (3.1)$$

with

$$\mathcal{L}(x^\alpha; \dot{x}^\alpha) = \left(\dot{x}^1\right)^2 + \left(\dot{x}^2\right)^2, \qquad (3.2)$$

where the dot denotes differentiation with respect to τ,

$$\dot{x}^\alpha = \frac{dx^\alpha}{d\tau}.$$

The problem of finding the shortest path between the two points has now been reduced to the problem of minimising $S[x^\alpha, \dot{x}^\alpha]$ under variations of the path $x^\alpha(\tau)$, keeping the end points fixed. This is a familiar problem. Under an infinitesimal such variation $x^\alpha(\tau) \to x^\alpha(\tau) + \delta x^\alpha(\tau)$, the variation of $S[x^\alpha, \dot{x}^\alpha]$ is

$$\begin{aligned}
\delta S[x^\alpha, \dot{x}^\alpha] &= \int_{\tau_0}^{\tau_1} \left(\frac{\delta \mathcal{L}}{\delta x^\alpha} \delta x^\alpha + \frac{\delta \mathcal{L}}{\delta \dot{x}^\alpha} \delta \dot{x}^\alpha \right) d\tau \\
&= \int_{\tau_0}^{\tau_1} \left(\frac{\delta \mathcal{L}}{\delta x^\alpha} \delta x^\alpha + \frac{\delta \mathcal{L}}{\delta \dot{x}^\alpha} \frac{d(\delta x^\alpha)}{d\tau} \right) d\tau \\
&= \int_{\tau_0}^{\tau_1} \left(\frac{\delta \mathcal{L}}{\delta x^\alpha} - \frac{d}{d\tau} \left(\frac{\delta \mathcal{L}}{\delta \dot{x}^\alpha} \right) \right) \delta x^\alpha d\tau,
\end{aligned}$$

where we have assumed that $\delta\left(\frac{dx^\alpha}{d\tau}\right) = \frac{d(\delta x^\alpha)}{d\tau}$ and integrated by parts in the last line, with the end points fixed, $\delta \mathbf{x}_0 = \delta \mathbf{x}_1 = 0$.

An extremum of $S[x^\alpha, \dot{x}^\alpha]$ is then obtained from

$$\delta S[x^\alpha, \dot{x}^\alpha] = 0 \quad \forall\, \delta x^\alpha \qquad \Leftrightarrow \qquad \frac{d}{d\tau} \left(\frac{\partial \mathcal{L}}{\partial \dot{x}^\alpha} \right) = \frac{\partial \mathcal{L}}{\partial x^\alpha}.$$

For (3.2) we have $\frac{\partial \mathcal{L}}{\partial x^\alpha} = 0$ and

$$\frac{\partial \mathcal{L}}{\partial \dot{x}^\alpha} = 2\dot{x}^\alpha = const.,$$

so the x^α are linear in τ. For simplicity, choose $\tau_0 = 0$; then the unique solution with fixed end points $\mathbf{x}(0) = \mathbf{x}_0$ and $\mathbf{x}(\tau_1) = \mathbf{x}_1$ is

$$x^1(\tau) = x_0^1 + (x_1^1 - x_0^1)\frac{\tau}{\tau_1}$$
$$x^2(\tau) = x_0^2 + (x_1^2 - x_0^2)\frac{\tau}{\tau_1}.$$

Strictly speaking, we still have to check whether this is a minimum or a maximum of (3.1), but it should be obvious that it is a minimum. We have proven that the shortest path between any two points in the plane (a geodesic) is a straight line.

Note that, if we multiply (3.2) by $\frac{m}{2}$ and interpret τ as Newtonian time, we can view (3.2) as the Lagrangian for a free particle of mass m and we have recovered Newton's first law: all particles not experiencing a force travel in straight lines with constant velocity.

3.2 Spherical Geometry

Now we shall determine the shortest path between two fixed points (θ_0, ϕ_0) and (θ_1, ϕ_1) on a sphere of constant radius a,

$$ds^2 = a^2(d\theta^2 + \sin^2\theta d\phi^2).$$

This is the problem of the airline pilots flying from India to the United States, shown in Figure 1.3.

We can always rotate our axes relative to the sphere to choose coordinates so that both points lie at the same value of θ, with $\theta_0 = \theta_1 = \frac{\pi}{2}$. There is no loss of generality in taking the two points to be at $(\theta_0, \phi_0) = (\frac{\pi}{2}, 0)$ and $(\theta_1, \phi_1) = (\frac{\pi}{2}, \phi_1)$ with $0 < \phi_1 \leq \pi$. Then parameterise the path by $\theta(\tau)$ and $\phi(\tau)$ with $\tau = 0$ at $(\frac{\pi}{2}, 0)$: and $\tau = \tau_1$ at $(\frac{\pi}{2}, \phi_1)$. Now

$$S[\dot\theta, \dot\phi; \theta, \phi] = \int_{\tau_0}^{\tau_1} \mathcal{L}(\dot\theta, \dot\phi; \theta, \phi) d\tau$$

with

$$\mathcal{L}(\dot\theta, \dot\phi; \theta, \phi) = a^2(\dot\theta^2 + \sin^2\theta\dot\phi^2),$$

where a dot denotes differentiation with respect to τ. The Lagrangian equations of motion are

$$\frac{d}{d\tau}\left(\frac{\partial\mathcal{L}}{\partial\dot\theta}\right) = \frac{\partial\mathcal{L}}{\partial\theta}, \qquad \frac{d}{d\tau}\left(\frac{\partial\mathcal{L}}{\partial\dot\phi}\right) = \frac{\partial\mathcal{L}}{\partial\phi},$$

giving

$$\ddot\theta = (\sin\theta\cos\theta)\dot\phi^2, \qquad \frac{d}{dt}\left(\sin^2\theta\dot\phi\right) = 0.$$

First integrals are immediate,

$$\dot\phi = \frac{A}{\sin^2\theta}, \qquad \dot\theta = -\frac{A^2}{2\sin^2\theta} + B,$$

where A and B are constants.

A solution, compatible with the preceding boundary conditions, is to set $\theta = \frac{\pi}{2}$ constant and $B = -A^2/2$, giving

$$\phi = A\tau,$$

with $A = \frac{\phi_1}{\tau_1}$.

This is not the only solution, though. We could set $A = \frac{\phi_1 + 2n\pi}{\tau_1}$ for any integer n, positive or negative, and this is also a solution. But for n non-zero this corresponds to winding round the sphere multiple times. These are paths that extremise the length, but the shortest path is $n = 0$.

We can put this into words and say that the shortest path between any two points on a sphere is a segment of a circle passing through the two points and whose centre coincides with the centre of the sphere. Such a circle necessarily has radius a and is called a *great circle* – it is a circle around the sphere with the greatest possible radius, the same radius as that of the sphere itself.

3.3 Minkowski Space-Time

In Minkowski space-time we can use the same approach to find geodesics, but we have to be a little bit careful because ds^2 in (2.16) might not be positive. Physically the most interesting geodesics are those corresponding to trajectories that are everywhere time-like (or light-like), as these can be the paths of real objects (or beams of light). For trajectories that are time-like at every point along their length, the initial and final points must also have a time-like separation and we restrict our considerations to paths with $ds^2 < 0$ everywhere along the path $(t(\tau), \mathbf{x}(\tau))$. Such a path is sometimes called the *world line* of the object, to emphasise the fact that time must flow – it is not possible for an object to sit at a point in space-time, an event, and not 'move' in time.

Extremal paths between two events, $(t_0, \mathbf{x}_0)$ and $(t_1, \mathbf{x}_1)$, are found by looking for extrema of

$$S[t, \mathbf{x}; \dot{t}, \dot{\mathbf{x}}] = \int_{t_0, \mathbf{x}_0}^{t_1, \mathbf{x}_1} \mathcal{L}(t, \mathbf{x}; \dot{t}, \dot{\mathbf{x}}) d\tau, \tag{3.3}$$

with

$$\mathcal{L}(t, \mathbf{x}; \dot{t}, \dot{\mathbf{x}}) = \dot{x}^2 + \dot{y}^2 + \dot{z}^2 - c^2 \dot{t}^2 < 0. \tag{3.4}$$

Our variational principle now essentially becomes the Lagrangian formulation of mechanics with 'time' τ. The equations of motion are

$$\ddot{t} = \ddot{\mathbf{x}} = 0, \tag{3.5}$$

and the solution is that $t(\tau)$ and $\mathbf{x}(\tau)$ are linear in τ.

We can always choose our coordinates so that $t_0 = 0$ and choose the spatial origin so that $\mathbf{x}_0 = 0$; then the unique solution is

$$t(\tau) = t_1 \frac{\tau}{\tau_1}, \qquad \mathbf{x}(\tau) = \mathbf{x}_1 \frac{\tau}{\tau_1},$$

where τ_1 is the value of τ at $(t_1, \mathbf{x}_1)$ – straight-line trajectories.

Putting the solution

$$t = \frac{t_1}{\tau_1}, \qquad \dot{\mathbf{x}} = \frac{\mathbf{x}_1}{\tau_1}$$

into (3.4) gives

$$\mathcal{L} = -c^2 \frac{(t_1^2 - \mathbf{x}_1 . \mathbf{x}_1)}{\tau_1^2} = -\frac{t_1^2}{\tau_1^2}(c^2 - v^2),$$

where $\mathbf{v} = \frac{d\mathbf{x}}{dt}$ is the velocity measured in the reference frame with inertial coordinates (ct, x, y, z).

If this is the trajectory of a body moving in space-time carrying its own internal clock, we shall interpret τ as the time shown on that clock – the body's proper time. We know from the special relativistic time-dilation effect that

$$dt = \gamma(v)d\tau \qquad \Rightarrow \qquad \frac{t_1}{\tau_1} = \gamma(v),$$

where $\gamma(v) = \frac{1}{\sqrt{1 - \frac{v^2}{c^2}}}$ is the Lorentz γ factor. We therefore have

$$\mathcal{L}(t, \mathbf{x}; \dot{t}, \dot{\mathbf{x}}) = -c^2,$$

when $\mathcal{L}$ is evaluated on the actual trajectory. In particular, when evaluated on the trajectory of a particle, $m\mathcal{L} = -mc^2$ is the rest energy of the particle.

When the solution for the trajectory is put into the action, we find

$$S[t, \mathbf{x}; \dot{t}, \dot{\mathbf{x}}] = -\tau_1 c^2 \qquad (3.6)$$

is proportional to the proper time τ_1 taken for the body to pass from event $(0, \mathbf{0})$ to event $(t_1, \mathbf{x}_1)$.

An alternative approach is, rather than fix the initial and final events, to fix the initial point and the initial velocity and leave the final point and the final value of τ free. Choose $\tau_0 = 0$ and initial conditions

$$t(0) = t_0, \quad \mathbf{x}(0) = \mathbf{x}_0, \qquad \dot{t}(0) = \gamma_0, \quad \dot{\mathbf{x}}(0) = \mathbf{u}_0,$$

with γ_0 and $\mathbf{u}_0$ constants. Solving (3.5) with these initial conditions leads to

$$t(\tau) = t_0 + \gamma_0 \tau, \qquad \mathbf{x}(\tau) = \mathbf{x}_0 + \mathbf{u}_0 \tau. \qquad (3.7)$$

Then

$$\mathbf{v} = \frac{d\mathbf{x}}{dt} = \frac{1}{\gamma_0}\mathbf{u}_0$$

and the trajectory is a straight line with constant velocity $\mathbf{v}_0 = \frac{1}{\gamma_0}\mathbf{u}_0$. If τ is the proper time along the trajectory, then $\gamma_0 = 1/\sqrt{1 - \frac{v_0^2}{c^2}}$ is the time-dilation factor.

Note that minimising S in (3.6) *maximises* the proper time τ_1, and this can have counter-intuitive consequences, such as the famous twin paradox in special relativity. Consider twins, Alice and Bob, one of whom (amazing Alice) goes off in a spaceship travelling at relativistic speeds to boldly go to a distant star, and then comes back again, while boring Bob stays at home on Earth. In the approximation in which the Earth is considered stationary, Bob is in an inertial reference frame and we can use his inertial coordinates to describe the trajectories. For simplicity, choose coordinates in which Alice moves in the x-direction and suppress y and z. Furthermore, choose the coordinates so that Bob sits at the origin and Alice leaves at $t = 0$ and returns at $t = t_B$ according to Bob's watch. Bob's trajectory is $t = \tau$, $x = 0$ in these coordinates and Alice returns at Bob's proper time $\tau_B = t_B$. For simplicity, suppose Alice travels at constant speed v to a star a distance d away from Bob and then immediately turns round and comes back at the same speed. From symmetry she will turn round at $t = t_B/2$.

On the outward journey, her trajectory is a straight line in the x-t plane, from (3.7)

$$t = \gamma(v)\tau, \quad x = \gamma(v)v\tau, \qquad 0 \le \tau \le \frac{t_B}{2\gamma(v)}, \quad d = \frac{vt_B}{2},$$

while on the return journey

$$t = \gamma(v)\tau, \quad x = d - v\gamma(v)\left(\tau - \frac{t_B}{2\gamma(v)}\right), \qquad \frac{t_B}{2\gamma(v)} \le \tau \le \frac{t_B}{\gamma(v)}.$$

The time on Alice's watch when she is at the farthest point away from Bob is $\frac{t_B}{2\gamma(v)}$; and, from symmetry, the time on her watch, her proper time, when she returns to Bob is

$$\tau_A = \frac{t_B}{\gamma(v)} = \frac{\tau_B}{\gamma(v)}.$$

If v is close to c, the time shown on Alice's watch is much less than that on Bob's; Bob has aged relative to Alice. The twin 'paradox' is that from Alice's point of view she is stationary and Bob has gone away and come back, so the same logic should imply that Bob returns younger than Alice, and both cannot be correct. But Alice and Bob are not on the same footing; Bob is in an inertial reference frame, and Alice is not.

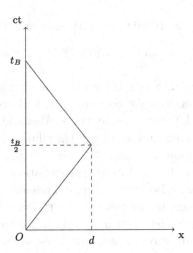

Figure 3.1 **The twin paradox.** Bob stays at home and follows the vertical trajectory, Alice (the two diagonal world lines) goes away and comes back. She cannot be in an inertial reference frame for the whole journey; she must accelerate. Their roles are not symmetric.

While it is true that, since v is constant, Alice is in an inertial reference frame on the way out and on the way back, she *must* accelerate when she turns round; she cannot be in an inertial reference frame. It is fallacious to say that she is 'almost' in an inertial reference frame since she only accelerates at one point.[2] For comparison, imagine Figure 3.1 with a 2-dimensional Euclidean metric: obviously, the sum of the lengths of the two diagonal sides would be greater than the length of the vertical side in a Euclidean metric; it would be nonsense to say that the two diagonal sides are a close approximation to a straight line because there is only a kink at one point. Similarly, in the Minkowski metric it would be nonsense to say that the two diagonal lines are a close approximation to a geodesic.

3.3.1 Rotating Coordinates

An interesting exercise is to take the Minkowski space-time line element (2.16) and transform to a rotating coordinate system,

$$\tilde{x} = x \cos \Omega t + y \sin \Omega t, \quad \tilde{y} = -x \sin \Omega t + y \cos \Omega t.$$

---

[2] Of course, the acceleration would be infinite if she really turned around at a point, but we can always round the point off a bit and make her trajectory smooth. This will complicate the mathematical analysis a bit, but the conclusion is the same.

Conversely, we can express (x, y) in terms of $(\tilde{x}, \tilde{y})$,

$$x = \tilde{x}\cos\Omega t - \tilde{y}\sin\Omega t, \quad y = \tilde{x}\sin\Omega t + \tilde{y}\cos\Omega t \tag{3.8}$$

and, of course,

$$x^2 + y^2 = \tilde{x}^2 + \tilde{y}^2.$$

Under an infinitesimal variation, (3.8) give

$$dx = (\cos\Omega t)d\tilde{x} - (\sin\Omega t)d\tilde{y} - \Omega(\tilde{x}\sin\Omega t + \tilde{y}\cos\Omega t)dt$$
$$dy = (\sin\Omega t)d\tilde{x} + (\cos\Omega t)d\tilde{y} + \Omega(\tilde{x}\cos\Omega t - \tilde{y}\sin\Omega t)dt$$
$$\Rightarrow \quad dx = (\cos\Omega t)d\tilde{x} - (\sin\Omega t)d\tilde{y} - \Omega y\,dt$$
$$dy = (\sin\Omega t)d\tilde{x} + (\cos\Omega t)d\tilde{y} + \Omega x\,dt,$$

from which we get

$$dx^2 + dy^2 = d\tilde{x}^2 + d\tilde{y}^2 + \Omega^2(\tilde{x}^2 + \tilde{y}^2)dt^2 + 2\Omega(\tilde{x}d\tilde{y} - \tilde{y}d\tilde{x})dt.$$

In this new coordinate system the Lagrangian (3.4) becomes

$$\mathcal{L} = \dot{\tilde{x}}^2 + \dot{\tilde{y}}^2 + \dot{z}^2 + 2\Omega(\tilde{x}\dot{\tilde{y}} + \tilde{y}\dot{\tilde{x}})\dot{t} - \{c^2 - \Omega^2(\tilde{x}^2 + \tilde{y}^2)\}\dot{t}^2.$$

The $z(\tau)$ equation is trivial, so let us set $z = const.$ and just consider motion in the x-y plane. The coordinate t is cyclic, the Lagrangian only depends on $\dot{t}$, and there is no explicit t dependence. We can thus immediately integrate the equation of motion for t to get

$$\dot{t} = \frac{kc^2 + \Omega(\tilde{x}\dot{\tilde{y}} - \tilde{y}\dot{\tilde{x}})}{c^2 - \Omega^2(\tilde{x}^2 + \tilde{y}^2)}, \tag{3.9}$$

where k is a dimensionless constant.

The $x(\tau)$ equation of motion follows from

$$\frac{\partial\mathcal{L}}{\partial\dot{\tilde{x}}} = 2\dot{\tilde{x}} - 2\Omega\tilde{y}\dot{t}, \qquad \frac{\partial\mathcal{L}}{\partial\tilde{x}} = 2\Omega^2\tilde{x}\dot{t}^2 + 2\Omega\dot{\tilde{y}}\dot{t},$$

from which

$$\frac{d}{d\tau}\left(\frac{\partial\mathcal{L}}{\partial\dot{\tilde{x}}}\right) = \frac{\partial\mathcal{L}}{\partial\tilde{x}}$$
$$\Rightarrow \quad \ddot{\tilde{x}} = \Omega\tilde{y}\ddot{t} + \Omega^2\tilde{x}\dot{t}^2 + 2\Omega\dot{\tilde{y}}\dot{t}. \tag{3.10}$$

Similarly, the $\tilde{y}$ equation of motion is

$$\ddot{\tilde{y}} = -\Omega\tilde{x}\ddot{t} + \Omega^2\tilde{y}\dot{t}^2 - 2\Omega\dot{\tilde{x}}\dot{t}. \tag{3.11}$$

Now take the non-relativistic limit, $c \to \infty$, and from (3.9) we see that $\dot{t} \to k$, $\ddot{t} \to 0$. So, we can set $\dot{\tilde{x}} = k\frac{dx}{dt}$, $\ddot{\tilde{x}} = k^2\frac{d^2x}{dt^2}$, $\dot{\tilde{y}} = k\frac{dy}{dt}$, and $\ddot{\tilde{y}} = k^2\frac{d^2y}{dt^2}$ and the equations of motion for $\tilde{x}$ and $\tilde{y}$ reduce to

$$\frac{d^2\tilde{x}}{dt^2} = \Omega^2\tilde{x} + 2\Omega\frac{d\tilde{y}}{dt}$$

$$\frac{d^2\tilde{y}}{dt^2} = \Omega^2\tilde{y} - 2\Omega\frac{d\tilde{x}}{dt}$$

or

$$\frac{d^2\tilde{\mathbf{r}}}{dt^2} = \Omega^2\tilde{\mathbf{r}} - 2\,\boldsymbol{\Omega} \times \frac{d\tilde{\mathbf{r}}}{dt}; \tag{3.12}$$

the first term is the centrifugal force and the second the Coriolis force.

Equation (3.9) looks problematic when $\Omega^2(\tilde{x}^2 + \tilde{y}^2) = c^2$; when the distance from the origin is large enough, the rotational velocity reaches the speed of light – but this is not a physical velocity; it is an artefact of the rotating coordinates system.

3.3.2 The Sagnac Effect

An interesting phenomenon in a rotating reference frame is the Sagnac effect. Consider a beam of light forced to move in a circle of radius a, for example round a fibre-optic cable, bent into a circle, centred on the origin in the x-y plane, and rotating round the z-axis with angular frequency Ω.

In cylindrical polar coordinates $x = \rho\cos\phi$, $y = \rho\sin\phi$,

$$dx^2 + dy^2 = d\rho^2 + \rho^2 d\phi^2$$

and the Minkowski line element (2.16) becomes

$$ds^2 = -c^2 dt^2 + d\rho^2 + \rho^2 d\phi^2 + dz^2.$$

Again focusing on motion in the x-y plane, we set $dz = 0$. Then, in a rotating reference frame with $\psi = \phi - \Omega t$ and $\rho = a$ fixed so that $d\rho = 0$,

$$ds^2 = -c^2\left(1 - \frac{\Omega^2 a^2}{c^2}\right)dt^2 + a^2 d\psi^2 + 2a^2\Omega d\psi dt.$$

The Lagrangian in the rotating reference frame is then

$$\mathcal{L}(t, \psi; \dot{t}, \dot{\psi}) = a^2\dot{\psi}^2 + 2a^2\Omega\dot{\psi}\dot{t} - (c^2 - a^2\Omega^2)\dot{t}^2.$$

A beam of light will follow a light-like trajectory, so $ds^2 = 0$ at every point along the trajectory and $\mathcal{L} = 0$ when evaluated on the trajectory. So we can take a shortcut and use

$$\mathcal{L} = a^2\dot{\psi}^2 + 2a^2\Omega\dot{\psi}\dot{t} - (c^2 - a^2\Omega^2)\dot{t}^2$$
$$= a^2(\dot{\psi} + \Omega\dot{t})^2 - c^2\dot{t}^2 = 0. \tag{3.13}$$

Solving (3.13), there are two solutions,

$$c\dot{t} = \pm a(\dot{\psi} + \Omega\dot{t}) \qquad \Leftrightarrow \qquad a\dot{\psi} = \mp(c \pm a\Omega)\dot{t}, \qquad (3.14)$$

corresponding to the light going clockwise or anti-clockwise around the loop. Hence, [3]

$$a\frac{d\psi}{dt} = a\frac{\dot{\psi}}{\dot{t}} = \mp(c \pm \Omega a).$$

It is as if the beam of light travelling in the same direction as the rotation is moving with speed $c - \Omega a$ and the beam travelling against the rotation is moving with speed $c + \Omega a$. This does not violate the tenets of relativity; the rotating reference frame is not an inertial reference frame – in an inertial reference frame the speed of light is c in both directions.

The light travelling with the rotation will take a time

$$t_+ = \frac{2\pi a}{c - \Omega a}$$

to go once round the loop; the light travelling against the rotation will take a shorter time

$$t_- = \frac{2\pi a}{c + \Omega a}$$

to go round once. This is perfectly natural; in the inertial reference frame in which the loop is rotating, the light travelling with the rotation has to travel farther to get back to its point of departure than the light travelling against the direction of rotation.

The time difference

$$\Delta t = t_+ - t_- = \frac{4\pi\Omega a^2}{c^2 - \Omega^2 a^2}$$

manifests itself as a phase shift between the two beams of light which can be measured using an interferometer. If the light has frequency ν and wavelength $\lambda = \frac{c}{\nu}$, then the angular frequency is

$$\omega = 2\pi\nu = \frac{2\pi c}{\lambda}.$$

---

[3] The parameter τ cannot be interpreted as proper time for a beam of light – a beam of light has no proper time.

$$\Delta\tau = \sqrt{1 - \frac{v^2}{c^2}}\Delta t \xrightarrow[v \to c]{} 0$$

for any Δt, no matter how large. $\Delta\tau$ always vanishes for a beam of light. In fact, $t(\tau)$ can be an arbitrary differentiable monotonically increasing function in the solution, and τ is just a mathematical parameter with no physical meaning.

The phase shift $\Delta\varphi$ (not to be confused with ϕ) is

$$\Delta\varphi = \omega\Delta t = \frac{8\Omega\pi^2 a^2 c}{\lambda(c^2 - \Omega^2 a^2)} \approx \frac{8\Omega\pi^2 a^2}{\lambda c},$$

where the last approximation is for $c \gg |\Omega|a$. The phase shift is small, because c is large, but interferometers are very good at measuring small phase shifts, and the Sagnac effect can be used to detect rotations – a fibre-optic gyroscope is extremely sensitive to rotations.

In fact, a similar formula works for a loop of any shape and area A:

$$\Delta\varphi = \frac{8\pi\Omega A}{\lambda c}.$$

3.4 Relativistic Particle Motion on a Sphere

Our next example is a 3-dimensional space-time with two space dimensions. Consider the Minkowski space-time line element (2.16) in spherical polar coordinates:

$$ds^2 = dr^2 + r^2(d\theta^2 + \sin^2\theta d\phi^2) - c^2 dt^2. \tag{3.15}$$

If we fix $r = a$ to be constant, this reduces to

$$ds^2 = a^2(d\theta^2 + \sin^2\theta d\phi^2) - c^2 dt^2,$$

and this represents a 3-dimensional space-time in which 2-dimensional space (a sphere of radius a) is curved. Suppose a particle follows a trajectory $(t(\tau), \theta(\tau), \phi(\tau))$ parameterised by τ, starting from an initial point $\theta = \theta_0$, $\phi = \phi_0$ on the sphere at $t = 0$, when $\tau = 0$, and moving with initial speed v_0 in an arbitrary direction. We can orient the sphere so that the initial velocity is in the direction of increasing ϕ with $\theta_0 = \frac{\pi}{2}$, $\phi_0 = 0$, and $v_0 = a\left.\frac{d\phi}{dt}\right|_0 = a\dot\phi_0/\dot t_0$. We then have initial conditions $\dot\theta_0 = 0$ and $a\dot\phi_0 = v_0\dot t_0$. If we chose τ to be the proper time of the particle, then time-dilation implies that $\dot t_0 = \gamma(v_0)$ and

$$\dot\phi_0 = \frac{\gamma(v_0)v_0}{a},$$

where $\gamma(v_0) = 1/\sqrt{1 - \frac{v_0^2}{c^2}}$ is the Lorentz γ-factor.

We can obtain the explicit form of the particle's subsequent world line by extremising

$$S[t, \theta, \phi; \dot t, \dot\theta, \dot\phi] = \int_0^\tau \mathcal{L}(t, \theta, \phi; \dot t, \dot\theta, \dot\phi) d\tau,$$

with

$$\mathcal{L}(t, \theta, \phi; \dot{t}, \dot{\theta}, \dot{\phi}) = a^2\left(\dot{\theta}^2 + \sin^2\theta\,\dot{\phi}^2\right) - c^2\dot{t}^2. \qquad (3.16)$$

The equations of motion are

$$c^2\ddot{t} = 0, \qquad \ddot{\theta} = \cos\theta\sin\theta\,\dot{\phi}^2, \qquad \frac{d}{d\tau}\left(\sin^2\theta\,\dot{\phi}\right) = 0,$$

from which

$$t(\tau) = k\tau, \qquad \dot{\phi} = \frac{l}{\sin^2\theta}, \qquad \dot{\theta} = -\frac{l^2}{2\sin^2\theta} + \tilde{l},$$

where k, l, and $\tilde{l}$ are constant.

One initial condition is that $k = \gamma(v_0)$. As in §3.2, where we studied geodesics on the sphere, we can set $\theta = \frac{\pi}{2}$ and $\tilde{l} = -l^2/2$, giving

$$\phi(\tau) = l\tau.$$

The initial conditions fix $l = \frac{\gamma(v_0)v_0}{a}$:

$$\phi(\tau) = \frac{\gamma(v_0)v_0\tau}{a},$$

and the final solution is that the particle travels on a great circle with constant speed

$$v = a\frac{d\phi}{dt} = a\frac{d\phi}{d\tau}\frac{d\tau}{dt} = v_0.$$

Putting the solution back into (3.16) gives

$$\mathcal{L} = -\gamma^2(v_0)\left(c^2 - v_0^2\right) = -c^2,$$

and

$$L = -c^2\tau$$

is proportional to the proper time τ; minimising L is equivalent to maximising the proper time. The particle takes the trajectory that maximises its proper time.

3.5 Newtonian Gravity

Newton's universal law of gravitation is closely related to the gravitational red-shift described in §2.7. If we take a 4-dimensional line element in which space is flat 3-dimensional Euclidean space but there is a position-dependent flow of time,

$$d^2s = -\left(1 + V(\mathbf{x})\right)c^2dt^2 + \delta_{\alpha\beta}dx^\alpha dx^\beta, \qquad (3.17)$$

where $\mathbf{x} = (x^1, x^2, x^3)$. Then the corresponding Lagrangian for a mass moving in this space-time is

$$\mathcal{L}(t, \mathbf{x}; \dot{t}, \dot{\mathbf{x}}) = \delta_{\alpha\beta}\dot{x}^\alpha\dot{x}^\beta - c^2(1 + V(\mathbf{x}))\dot{t}^2.$$

The variational principle gives

$$\ddot{x}^\alpha = -\frac{c^2}{2}(\partial_\alpha V)\dot{t}^2, \qquad \frac{\partial}{\partial\tau}\left\{(1 + V(\mathbf{x}))\dot{t}\right\} = 0.$$

So

$$\dot{t} = \frac{k}{1 + V}$$

where k is a constant and

$$\ddot{x}^\alpha = -\frac{k^2 c^2}{2(1 + V)^2}\partial_\alpha V.$$

We can choose $k = 1$ and define $\Phi(\mathbf{r}) = \frac{c^2}{2}V(\mathbf{x})$; then the acceleration is

$$\ddot{\mathbf{x}} = -\frac{1}{(1 + \frac{2}{c^2}\Phi)^2}\nabla\Phi.$$

In the non-relativistic limit $c \to \infty$, keeping c^2V finite so $V \to 0$ and $\dot{t} = 1$, the acceleration is

$$\mathbf{a} = \ddot{\mathbf{x}} = -\nabla\Phi.$$

In the non-relativistic limit, geodesics of this metric correspond to accelerations in a gravitational field generated by the Newtonian gravitational potential Φ, as in (1.7).

Note that, if the gravitational potential Φ is generated by a mass M centred on the origin, $\Phi = -\frac{GM}{r}$, then $V = \frac{GM}{c^2 r}$ and the line element (3.17) is

$$d^2 s = -\left(1 - \frac{2GM}{c^2 r}\right)c^2 dt^2 + dx^\alpha dx^\alpha,$$

which is the non-relativistic (large c) limit of (2.31).

3.6 Cosmological Geometry and the Expansion of the Universe

In this section we explore geodesics in an expanding universe (2.21). To find geodesics we use τ as a parameter along the trajectory and vary $x(\tau)$, $y(\tau)$, $z(\tau)$, and $t(\tau)$ in the corresponding Lagrangian:

$$\mathcal{L}(t, \mathbf{x}; \dot{t}, \dot{\mathbf{x}}) = a^2(t)(\dot{x}^2 + \dot{y}^2 + \dot{z}^2) - c^2\dot{t}^2. \tag{3.18}$$

The equations of motion are

$$\frac{d}{d\tau}(c^2 \dot{t}) = -a\frac{da}{dt}(\dot{x}^2 + \dot{y}^2 + \dot{z}^2) = -a\frac{da}{d\tau}\frac{d\tau}{dt}(\dot{x}^2 + \dot{y}^2 + \dot{z}^2),$$
$$\frac{d}{d\tau}(a^2 \dot{x}) = \frac{d}{d\tau}(a^2 \dot{y}) = \frac{d}{d\tau}(a^2 \dot{z}) = 0.$$

In particular, $a^2 \dot{x}$, $a^2 \dot{y}$, and $a^2 \dot{z}$ are constant, so $u_0^2 = a^4(\dot{x}^2 + \dot{y}^2 + \dot{z}^2)$ is constant, and

$$\frac{1}{2}\frac{d}{d\tau}(\dot{t}^2) = -\frac{u_0^2}{c^2 a^3}\frac{da}{d\tau} = \frac{1}{2}\frac{d}{d\tau}\left(\frac{u_0^2}{c^2 a^2}\right)$$

$$\Rightarrow \quad \dot{t}^2 - \frac{u_0^2}{c^2 a^2} = const.$$

From (2.22) the initial speed is $v_0 = a(t_0)\frac{dx}{dt}\big|_0$,

$$v_0 = \frac{a_0\sqrt{\dot{x}_0^2 + \dot{y}_0^2 + \dot{z}_0^2}}{\dot{t}_0} = \frac{u_0}{a_0 \dot{t}_0},$$

and we take τ to be the body's proper time; so $t_0 = \gamma(v_0)\tau_0$ and $\dot{t}_0 = \gamma(v_0)$, so

$$v_0 = \frac{u_0}{a_0\gamma(v_0)} \qquad (3.19)$$

and initially

$$\dot{t}_0^2 - \frac{u_0^2}{c^2 a_0^2} = \dot{t}_0^2 - \frac{\gamma^2(v_0)v_0^2}{c^2} = \gamma^2(v_0)\left(1 - \frac{v_0^2}{c^2}\right) = 1,$$

where $\dot{t}_0$ and a_0 are the initial values of $\dot{t}$ and a, respectively.

This gives us, using (3.19),

$$\dot{t}^2 = 1 + \frac{u_0^2}{c^2 a^2} = 1 + \gamma^2(v_0)\frac{v_0^2}{c^2}\left(\frac{a_0}{a}\right)^2.$$

Using (2.22) again, the speed at any time $t > t_0$ is given by

$$v(\tau) = \frac{a\sqrt{\dot{x}^2 + \dot{y}^2 + \dot{z}^2}}{\dot{t}}$$

and

$$\dot{x}^2 + \dot{y}^2 + \dot{z}^2 = \frac{u_0^2}{a^4},$$

so finally

$$v(t) = \frac{u_0}{a\sqrt{1 + \gamma^2(v_0)\frac{v_0^2}{c^2}\frac{a_0^2}{a^2}}} = \frac{\gamma(v_0)v_0}{\sqrt{\frac{a^2}{a_0^2} + \gamma^2(v_0)\frac{v_0^2}{c^2}}}. \qquad (3.20)$$

If $a(t)$ is an increasing function of t, then $v(t)$ is a monotonically decreasing function and, if $a(t)$ increases indefinitely, $v(t)$ tends to zero. This appears to be the case in our Universe – this is called *heat death*. If v is the speed of a gas particle at temperature T, the speed, and hence the temperature, decreases as a function of time and everything gets colder in the future until $T \to 0$ as $t \to \infty$. In the limit of small speeds, let $c \to \infty$ in (3.20) and

$$v(t) \to \frac{a_0}{a(t)} v_0;$$

the speed simply decreases as $\sim 1/a$.

In the opposite direction, if a is smaller for times $t < t_0$, v increases as t decreases, and if $a(t)$ ever tends to zero, then $v \to c$ and all matter in the Universe becomes relativistic and very hot – these were the conditions in the very early Universe just after the Big Bang.

3.7 The Schwarzschild Geometry and Black Holes

In this section we look at some geodesics in the Schwarzschild space-time metric (2.31). A test particle moving in this space-time will follow a trajectory $(ct(\tau), r(\tau), \theta(\tau), \phi(\tau))$, where τ parameterises events along the particle's world line. (It is convenient to choose τ to be the particle's proper time but this is not necessary.) The Lagrangian governing the particle's motion will be

$$\mathcal{L}(t, r, \theta, \phi; \dot{t}, \dot{r}, \dot{\theta}, \dot{\phi}) = \frac{\dot{r}^2}{\left(1 - \frac{2GM}{c^2 r}\right)} + r^2\left(\dot{\theta}^2 + \sin^2\theta\,\dot{\phi}^2\right) - \left(1 - \frac{2GM}{c^2 r}\right) c^2 \dot{t}^2.$$
$$(3.21)$$

Solving the equations of motion for this Lagrangian is the general relativistic version of the Kepler problem in Newtonian gravity. Note that the Lagrangian is symmetric under rotations about the spatial origin – this symmetry translates to conservation of angular momentum in the dynamics of a test particle.

The equations of motion are

$$\frac{d}{d\tau}\left\{\left(1 - \frac{2GM}{c^2 r}\right)\dot{t}\right\} = 0,$$

$$\frac{d}{d\tau}\left\{\frac{\dot{r}}{\left(1 - \frac{2GM}{c^2 r}\right)}\right\} = -\frac{GM\dot{t}^2}{r^2} - \frac{\dot{r}^2}{\left(1 - \frac{2GM}{c^2 r}\right)^2}\left(\frac{GM}{c^2 r^2}\right)$$
$$+ r\left(\dot{\theta}^2 + \sin^2\theta\,\dot{\phi}^2\right),$$

$$\frac{d}{d\tau}\left(r^2\dot\theta\right) = r^2\cos\theta\sin\theta\dot\phi^2,\qquad(3.22)$$

$$\frac{d}{d\tau}\left(r^2\sin^2\theta\,\dot\phi\right) = 0.$$

The last two equations are exactly the same as in the Kepler problem; conservation of angular momentum dictates that the motion of the test particle restricted to lie in a 2-dimensional plane, and without any loss of generality we can set $\theta = \frac{\pi}{2} = const.$, in which case the last equation then requires $r^2\dot\phi$ to be constant,

$$r^2\dot\phi = l \quad\Rightarrow\quad \dot\phi = \frac{l}{r^2},\qquad(3.23)$$

where l is the angular momentum per unit mass of the test particle.

The first equation in (3.22) implies that $\left(1 - \frac{2GM}{c^2r}\right)\dot t$ is a constant of the motion:

$$\left(1 - \frac{2GM}{c^2r(\tau)}\right)\dot t(\tau) = k.\qquad(3.24)$$

This constant of motion arises from the fact that $t(\tau)$ is a cyclic coordinate; the Lagrangian depends on $\dot t(\tau)$ but not explicitly on $t(\tau)$ itself, which in turn reflects the fact that the Lagrangian is symmetric under time translations. This means that energy is conserved: as we shall see, k is related to the energy of the test particle.

Equation (3.24) is a manifestation of gravitational red-shift: if $k = 1$, we can interpret the coordinate t to be the proper time of an observer fixed at $r = \infty$, $t_{r=\infty} = \tau_\infty$. As r decreases, $\dot t = \frac{k}{1-\frac{2GM}{rc^2}}$ increases and the proper time τ for a clock at finite fixed r is less than t; time slows down in a gravitational field.

Lastly, employing (3.23) and (3.24), the equation of motion for $\dot r$ reduces to

$$\ddot r = -\frac{GM}{r^2}\frac{k^2}{\left(1-\frac{2GM}{c^2r}\right)} + \frac{\dot r^2}{\left(1-\frac{2GM}{c^2r}\right)}\left(\frac{GM}{c^2r^2}\right) + \left(1-\frac{2GM}{c^2r}\right)\frac{l^2}{r^3}.\qquad(3.25)$$

In the non-relativistic limit, $c \to \infty$, we can set $t = \tau$, so $k = 1$ and

$$\ddot r = -\frac{GM}{r^2} + \frac{l^2}{r^3}\qquad(3.26)$$

is the orbit equation for a planet moving around the Sun, where M is the mass of the Sun and the second term on the right-hand side is the centrifugal force arising from the planet's orbital motion. We know that energy is conserved, the non-relativistic energy per unit mass of the test particle is

$$E = \frac{\dot{r}^2}{2} - \frac{GM}{r} + \frac{l^2}{2r^2}, \qquad (3.27)$$

and (3.26) follows from

$$\frac{dE}{d\tau} = 0.$$

We can try to find a first integral of (3.25) directly, but a neater way is to appeal to the discussion on page 48 and interpret $-m\mathcal{L}$ as being the rest energy of the test particle when τ is chosen to be the proper time. Thus, $\mathcal{L} = -c^2$ when evaluated on the geodesic, and putting (3.23) and (3.24) into (3.21) directly results in

$$\boxed{c^2(k^2 - 1) = \dot{r}^2 - \frac{2GM}{r} + \left(1 - \frac{2GM}{c^2 r}\right)\frac{l^2}{r^2}.} \qquad (3.28)$$

It is left as an exercise to show that differentiating this equation with respect to τ reproduces the relativistic orbit equation (3.25), when k^2 in (3.25) is eliminated using (3.28) itself.

Setting $k^2 - 1 = 2E/c^2$ in (3.28), we see that it is very similar to the non-relativistic equation (3.27) and indeed only differs from it by the addition of the term $-\frac{2GMl^2}{c^2 r^3}$, which vanishes in the non-relativistic limit $c \to \infty$.

We can try to find solutions of the general relativistic orbit equation using techniques familiar from the Kepler problem. As for the Kepler problem, it is useful to re-express the orbit equation in terms of $1/r$ rather than r itself. Let

$$u = \frac{r_S}{r} \qquad \text{with} \qquad r_S = \frac{2GM}{c^2}, \qquad (3.29)$$

the Schwarzschild radius, so u is dimensionless. (For the mass of the Sun r_S is 3 km, so u is very small in the dynamics of the Solar System.) Then,

$$\dot{r} = \frac{dr}{d\phi}\frac{d\phi}{d\tau} = \frac{dr}{d\phi}\frac{l}{r^2} = -\frac{r_S}{u^2}\frac{du}{d\phi}\frac{lu^2}{r_S^2} = -\frac{l}{r_S}\frac{du}{d\phi}.$$

Instead of trying to solve (3.28), for $r(\tau)$, we write it as

$$\left(\frac{du}{d\phi}\right)^2 = \left(\frac{cr_S}{l}\right)^2 (k^2 - 1 + u) - u^2(1 - u) \qquad (3.30)$$

and solve for $u(\phi)$. Differentiating this equation with respect to ϕ results in a second-order ordinary differential equation for $u(\phi)$,

$$\boxed{\frac{d^2u}{d\phi^2} = \frac{1}{2}\left(\frac{cr_S}{l}\right)^2 - u + \frac{3}{2}u^2.} \qquad (3.31)$$

This is the general relativistic orbit equation; the $\frac{3}{2}u^2$ term represents the general relativistic correction to the Newtonian orbit equation. This is not a linear equation, and we shall analyse the solutions using perturbations about the non-relativistic Newtonian solution.

3.7.1 Precession of Perihelion

If u is small, we can try to solve (3.31) approximately, since then u^2 will be small relative to u. As a first approximation, ignore the u^2 term and consider

$$\frac{d^2u}{d\phi^2} = \frac{1}{2}\left(\frac{cr_S}{l}\right)^2 - u.$$

This is the non-relativistic Newtonian orbit equation in disguise; in ignoring u^2 we are ignoring relativistic corrections. As for the familiar Kepler problem, this is just the harmonic oscillator equation and the general solution is immediate:

$$u(\phi) - \frac{1}{2}\left(\frac{cr_S}{l}\right)^2 = A\cos(\phi + \delta),$$

where A and δ are arbitrary integration constants. We might as well set $\delta = 0$, as it just corresponds to different choices of the $\phi = 0$ direction:

$$u(\phi) = \frac{1}{2}\left(\frac{cr_S}{l}\right)^2 + A\cos\phi.$$

In terms of $r(\phi)$, this reads

$$r(\phi) = \frac{r_0}{(1 + e\cos\phi)}, \tag{3.32}$$

where

$$e = \frac{l^2c^2A}{2G^2M^2} \quad \text{and} \quad r_0 = \frac{l^2}{GM}.$$

This is Kepler's first law: for $0 < e < 1$, (3.32) is the equation of an ellipse with a focus at the origin, e is the eccentricity, and r_0 a parameter, called the *semi-latus rectum*, that sets the size of the ellipse (see problem 3 in Chapter 2). The ellipse keeps a constant orientation in space; $r(\phi)$ is a minimum when $\cos(\phi) = 1$, that is, when ϕ is zero or an integer multiple of 2π; and $r(\phi)$ is a maximum when $\cos(\phi) = -1$. This is an essential feature of Newton's inverse square law; it would not be the case for inverse cube, for example.[4]

---

[4] The fact that the ellipse keeps a constant orientation in space is a consequence of symmetry. Any central force, directed towards the origin, gives rise to a potential energy $V(r)$ that is symmetric under rotations in 3-dimensional space, and this

So, if we ignore the u^2 term, a solution is

$$u(\phi) = \frac{r_S}{r(\phi)} = \frac{r_S}{r_0}(1 + e\cos\phi).$$

But this is only an approximate solution of (3.31) when u is small, so we shall define

$$\epsilon = \frac{r_S}{r_0} = 2\left(\frac{GM}{cl}\right)^2$$

and expand in ϵ to get a better solution.[5] The full solution can be written as

$$u(\phi) = \epsilon u_1(\phi) + \epsilon^2 u_2(\phi) + \epsilon^3 u_3(\phi) + \ldots, \tag{3.33}$$

which is an expansion in $1/c^2$, and we have just calculated

$$u_1(\phi) = 1 + e\cos\phi. \tag{3.34}$$

So, now we go to the next order in ϵ and calculate $u_2(\phi)$. Putting (3.34) in (3.33) and using this in (3.31), we get (ignoring terms of order ϵ^3) a linear inhomogeneous differential equation for $u_2(\phi)$,

$$\frac{d^2 u_2}{d\phi^2} + u_2 = \frac{3}{2}(1 + e\cos\phi)^2$$

$$= \frac{3}{4}(2 + e^2) + 3e\cos\phi + \frac{3e^2}{4}\cos(2\phi). \tag{3.35}$$

This is the equation of a forced harmonic oscillator, and it can be solved by standard techniques. The general solution of (3.35) is obtained by first finding the most general solution of the corresponding homogeneous equation

$$\frac{du_2}{d\phi^2} + u_2 = \frac{3}{4}(2 + e^2),$$

which is

$$u_2 = \frac{3}{4}(2 + e^2) + B\cos(\phi + \delta), \tag{3.36}$$

---

results in conservation of angular momentum, **L**. For the special case of an inverse square law, the potential is $V(r) = \frac{a}{r}$, where a is a constant, and the *Hamiltonion* $H(p,r) = \frac{p^2}{2m} + \frac{a}{r}$ enjoys a larger symmetry than other power law potentials would give. There is another conserved quantity associated with this symmetry, a vector $\mathbf{R} = \mathbf{p} \times \mathbf{L} + ma\hat{r}$, called the *Runge–Lenz* vector, which is constant. The fact that this vector keeps a constant orientation in space translates to an ellipse with constant orientation.

[5] For the Sun, $r_S = 2.95 \times 10^3$ m and $r_0 > r_\odot$, where $r_\odot = 6.98 \times 10^8$ m is the radius of the Sun. So, $\epsilon < 4.3 \times 10^{-6}$ at the surface of the Sun, and even less at the radius of any planetary orbit.

with B a constant, and then adding any particular solution of the full inhomogeneous equation (3.35). Now (3.36) is of the same form as u_1 and we can simply absorb it into ϵu_1 with a suitable shift in constants. A particular solution of (3.35) can then easily be found from experience with forced harmonic oscillators, with forcing term $3e\cos\phi - \frac{3e^2}{4}\cos(2\phi)$. Note that, from this perspective, the $3e\cos\phi$ term in (3.35) is forcing at the resonance frequency of the oscillator, and this will give rise to a very interesting result. A particular solution of (3.35) is

$$u_2 = \frac{3}{4}(2 + e^2) + \frac{3e}{2}\phi\sin\phi - \frac{e^2}{4}\cos(2\phi)$$

$$= \frac{3}{2} + e^2 + \frac{3e}{2}\phi\sin\phi - \frac{e^2}{2}\cos^2\phi.$$

Again the constant term can be absorbed into ϵu_1 and we are left with

$$u(\phi) = \epsilon(1 + e\cos\phi) + \frac{1}{2}\epsilon^2(3e\phi\sin\phi - e^2\cos^2\phi) + O(\epsilon^3),$$

hence

$$r(\phi) = \frac{r_0}{1 + e\cos\phi + \frac{\epsilon}{2}(3e\phi\sin\phi - e^2\cos^2\phi)} + O(\epsilon^2). \tag{3.37}$$

The term $3e\phi\sin\phi$ here is crucial; no matter how small ϵ is, ϕ increases indefinitely as the orbit winds around the origin, and eventually this term must have an observable effect. To quantify this, we first calculate the point of closest approach. For $r(\phi)$ to be a minimum, a necessary requirement is $\frac{dr(\phi)}{d\phi} = 0$. Evaluating the derivative up to order ϵ,

$$\frac{dr(\phi)}{d\phi} = -\frac{r_0\left(-e\sin\phi + \frac{\epsilon}{2}(3e\sin\phi + 3e\phi\cos\phi + 2e^2\sin\phi\cos\phi)\right)}{\left(1 + e\cos\phi + \frac{\epsilon}{2}(3e\phi\sin\phi - e^2\cos^2\phi)\right)^2} + O(\epsilon^2). \tag{3.38}$$

Clearly $r(\phi)$ is a minimum at $\phi = 0$, but the next minimum does not occur at $\phi = 2\pi$; it is at a slightly displaced value $\phi = 2\pi + \delta$. δ will vanish when $\epsilon = 0$, but when ϵ is non-zero, but small, δ will be non-zero and small. Indeed, we expect that δ will be of the same order of magnitude as ϵ, in which case

$$\sin(2\pi + \delta) = \delta + O(\epsilon^2) \quad \text{and} \quad \cos(2\pi + \delta) = \cos(\delta) \approx 1 - \frac{\delta^2}{2} = 1 + O(\epsilon^2).$$

Using these in (3.38) $\frac{dr}{d\phi} = 0$ gives the next minimum at $\phi = 2\pi + \delta$ when

$$e\sin\phi = \frac{3\epsilon e}{2}\phi\cos\phi + O(\epsilon^2) \quad \Rightarrow \quad e\delta + O(\epsilon^2) = 3\pi\epsilon e + O(\epsilon^2),$$

so

$$\delta = 3\pi\epsilon. \tag{3.39}$$

For a planet orbiting the Sun, the ellipse does not keep a constant orientation in space; perihelion advances by $3\pi\epsilon$ on each revolution. The effect is cumulative and continues to grow; after n revolutions,

$$\delta_n = 3\pi n\epsilon. \tag{3.40}$$

This is a consequence of the fact that (3.35) has the form of a forced harmonic oscillator at resonance.

The effect is largest for the planet with the largest value of $\epsilon = \frac{r_S}{r_0}$, and this requires the smallest r_0; this is Mercury, whose orbit has $r_0 = 55.46 \times 10^6$ km. For the Sun, the Schwarzschild radius is $r_S = 2,954$ km, so $\epsilon = 5.326 \times 10^{-8}$ for Mercury and

$$\delta_n = (5.020 \times 10^{-7}) \times n$$

(in radians). Although this is small, we have the advantage that it increases with each revolution, and grows the longer we wait. Mercury orbits the Sun once every 87.97 days, so in one hundred years it orbits 415 times and

$$\delta_{415} = 2.083 \times 10^{-4} \text{ radians} = 0.01194°, \tag{3.41}$$

which is small but measurable. The observed advance is actually $0.15947° \pm 0.00018°$ in 100 years, but $0.14767° \pm 0.00019°$ of this is calculated to be due to the Newtonian gravitational attraction of the other planets, mostly Venus and Jupiter; the remaining $0.01180° \pm 0.00019°$ cannot be accounted for by Newtonian gravity and agrees well with Einstein's general relativistic prediction (3.41).

3.7.2 The Bending of Light by the Sun

One of the most successful predictions of Einstein's general relativity when it was first conceived in 1915 is that the Sun will bend a beam of light that just grazes its edge, by an amount that was measured in 1919 during a total solar eclipse that was visible in the Gulf of Guinea off the west coast of Africa.

A beam of light in the gravitational field of the Sun will follow a trajectory described by a geodesic which is a solution of Equations (3.22). We can use the same analysis as in the previous section with the one difference that light, naturally enough, follows light-like rather than time-like geodesics. This means that the line element is $d^2s = 0$ and that $\mathcal{L}$ in (3.21), when evaluated on the trajectory, is not $-c^2$, but vanishes for light. (Another way of seeing this is that light is composed of massless particle, photons, so we set $mc^2 = 0$.) Working through the details, Equation (3.28) is replaced with

$$c^2 k^2 = \dot{r}^2 + \left(1 - \frac{2GM}{c^2 r}\right) \frac{l^2}{r^2}. \tag{3.42}$$

Again converting this from a differential equation for $r(\tau)$ to one for $u(\phi) = \frac{r_S}{r(\phi)}$, we find

$$\left(\frac{du}{d\phi}\right)^2 = \frac{r_S^2}{l^2} c^2 k^2 - (1-u)u^2. \tag{3.43}$$

Differentiating once more with respect to ϕ gives the second-order equation

$$\frac{d^2 u}{d\phi^2} = -u + \frac{3}{2}u^2,$$

which differs from (3.31) only by the absence of the constant term on the right-hand side. Again we can solve perturbatively; provided $r \gg r_S$ for the whole trajectory, then $u \ll 1$. Suppose the beam of light passes the Sun, with r_0 the point of closest approach. Let $\epsilon = \frac{r_S}{r_0} \ll 1$. Expand

$$u(\phi) = \epsilon u_1(\phi) + \epsilon^2 u_2(\phi) + \epsilon^3 u_3(\phi) + \dots. \tag{3.44}$$

Then, at first order in ϵ, we again have the harmonic oscillator equation,

$$\frac{d^2 u_1}{d\phi^2} + u_1 = 0,$$

with solution

$$u_1(\phi) = \sin\phi.$$

The amplitude is unity because the phase is chosen so that the point of closest approach is when u_1 is largest, that is, $\phi = \frac{\pi}{2}$ when $u = \frac{r_S}{r_0} = \epsilon \Rightarrow u_1 = 1$. At the next iteration, (3.44) in (3.43) gives

$$\frac{d^2 u_2}{d\phi^2} + u_2 = \frac{3}{2}u_1^2 = \frac{3}{4}\big(1 - \cos(2\phi)\big).$$

This is a forced harmonic oscillator equation with a particular solution

$$u_2 = \frac{1}{4}\big(3 + \cos(2\phi)\big),$$

giving

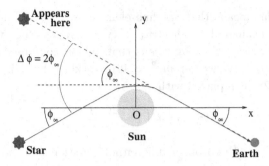

Figure 3.2 **Light bending by the gravitational field of the Sun.**
(The effect is greatly exaggerated for clarity.)

$$u(\phi) = \epsilon \sin \phi + \frac{\epsilon^2}{4}\left(3 + \cos(2\phi)\right) + O(\epsilon^3). \tag{3.45}$$

Suppose a beam of light comes in towards the Sun from $r = \infty$, where $u = 0$ and ϕ is very small, $|\phi_\infty| \ll 1$. Approximating $\sin \phi_\infty \approx \phi_\infty$ and $\cos(2\phi_\infty) \approx 1$, $u = 0$ in (3.45) implies

$$0 = \phi_\infty + \epsilon + O(\epsilon^2),$$

hence $\phi_\infty = -\epsilon$ and the magnitude of the total deflection (see Figure 3.2) is

$$|\Delta\phi| = 2\epsilon.$$

The greatest effect is if the beam of light just grazes the Sun's limb, with $r_0 = r_\odot = 6.95 \times 10^8$ m the radius of the Sun, as this minimises r_0, in which case

$$|\Delta\phi| = 2\frac{r_S}{r_\odot} = 8.5 \times 10^{-6} \text{ radians,}$$

or about $1/2{,}000$ of a degree.

While this is a tiny amount, it can be measured. We cannot see stars during the day; it is difficult to image stars close to the Sun, but we can make use of total solar eclipses. The idea is to find a group of stars in the night sky, which you know the Sun is going to be close to during a solar eclipse six months later, and photograph them at night when the Sun is nowhere near them. Wait for the eclipse, when the Sun is near the same group of stars, and photograph them during the eclipse. When the photographs are compared, stars very close to the Sun will be slightly displaced radially away from the Sun compared to their usual position, by $1/2{,}000$ of a degree, while stars a few degrees or so from the Sun will not be displaced appreciably.

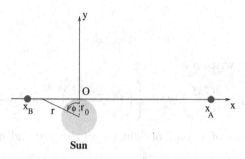

Figure 3.3 **The geometry of the Shapiro time delay measurement**.

Such an eclipse occurred on the 29th of May 1919, when the Sun was among a group of stars called the Hyades cluster in the constellation of Taurus. The eclipse was only visible close to the equator, from Brazil to Africa, and an expedition was arranged to go and observe it on the Island of Principe off the west coast of Africa, by the British astronomers Frank Watson Dyson and Arthur Eddington, and to photograph it. The expedition was reported as hugely successful; Einstein's prediction was verified to great acclaim in the public press.

3.7.3 Shapiro Time Delay

If a beam of light, or a radar pulse, passes close to the Sun, it takes slightly longer for its journey in the Schwarzschild geometry than it would in flat space-time. This is a measurable effect, most easily detected using radio waves, and is known as the *Shapiro time delay*. In the following we shall talk about 'light' for conciseness, but this should be understood to mean electromagnetic radiation of any wavelength; in real measurements radio waves are used. Consider a beam of light going from a point A to a point B, passing close to the Sun as shown in Figure 3.3. Choose coordinates so that the trajectory lies in the x-y plane ($z = 0$ or $\theta = \frac{\pi}{2}$) with the point of closest approach at $x = y = 0$.

With r defined as in the figure, the Schwarzschild line element in the x-y plane reads

$$ds^2 = -c^2 \left(1 - \frac{2GM}{c^2 r}\right) dt^2 + \left(1 - \frac{2GM}{c^2 r}\right)^{-1} dr^2 + r^2 d\phi^2. \quad (3.46)$$

If the point of closest approach to the Sun has radial coordinate r_0, then, from the figure,

$$x = -r \sin\phi = -r_0 \tan\phi, \qquad r_0 = r \cos\phi$$

and

$$dr = r\tan\phi\,d\phi,$$

so (3.46) can be written as[6]

$$ds^2 = -c^2\left(1 - \frac{2GM}{c^2 r}\right)dt^2 + \left\{\left(1 - \frac{2GM}{c^2 r}\right)^{-1}\tan^2\phi + 1\right\}r^2 d\phi^2.$$

For the trajectory of a beam of light we set $ds^2 = 0$ and find

$$c^2 dt^2 = \left(1 - \frac{2M}{c^2 r}\right)^{-2}\left(\tan^2\phi + 1 - \frac{2M}{c^2 r}\right)r^2 d\phi^2.$$

Now

$$d\phi = -\frac{\cos^2\phi}{r_0}dx \qquad \text{and} \qquad r\,d\phi = -\cos\phi\,dx,$$

giving

$$\begin{aligned}
c^2 dt^2 &= \left(1 - \frac{2M}{c^2 r}\right)^{-2}\left(1 - \frac{2M}{c^2 r}\cos^2\phi\right)dx^2 \\
&= \left(1 - \frac{r_0}{r}\epsilon\right)^{-2}\left(1 - \frac{r_0}{r}\epsilon\cos^2\phi\right)dx^2 \qquad (3.47) \\
&= \left(1 - \epsilon\cos\phi\right)^{-2}\left(1 - \epsilon\cos^3\phi\right)dx^2,
\end{aligned}$$

where, as before, $\epsilon = \frac{2GM}{c^2 r_0} < 4.3\times 10^{-6}$. For convenience we choose to work with the dimensionless variable

$$u = \tan\phi = -\frac{x}{r_0},$$

in terms of which

$$\cos\phi = \frac{1}{\sqrt{1 + u^2}},$$

so finally

$$dt = \frac{r_0}{c}\left(1 - \frac{\epsilon}{(1 + u^2)^{\frac{1}{2}}}\right)^{-1}\left(1 - \frac{\epsilon}{(1 + u^2)^{\frac{3}{2}}}\right)^{\frac{1}{2}}du \qquad (3.48)$$

and the total time taken for light to pass from A to B along this trajectory is

---

[6] We are assuming the trajectory is a straight line, but, of course, it is bent due to light deflection. But this bending is a higher-order effect that is not relevant for the time delay effect being discussed here. We are interested in the difference in the time it takes for light to pass from A to B when the Sun is in the way, compared to the time it takes when the Sun is somewhere else in the sky; the bending of the light gives a tiny correction to this time difference.

$$T = \frac{r_0}{c} \int_{u_A}^{u_B} \left(1 - \frac{\epsilon}{\left(1 + u^2\right)^{\frac{1}{2}}}\right)^{-1} \left(1 - \frac{\epsilon}{\left(1 + u^2\right)^{\frac{3}{2}}}\right)^{\frac{1}{2}} du. \quad (3.49)$$

The integral can be done numerically; for example, if the Earth is at A and Venus is in opposition on the opposite side of the Sun at B, so that the light beam just grazes the Sun's limb at $r_0 = 6.96 \times 10^8$ m, then $\epsilon = 4.24 \times 10^{-6}$ with[7]

$$u_A = -216 \quad \text{and} \quad u_B = 156. \quad (3.50)$$

However, since ϵ is so small, we can Taylor expand in ϵ, and keeping only the first term gives the answer to one part in a million,

$$T = \frac{r_0}{c} \int_{u_A}^{u_B} \left(1 + \frac{\epsilon}{\left(1 + u^2\right)^{\frac{1}{2}}} - \frac{\epsilon}{2\left(1 + u^2\right)^{\frac{3}{2}}} + O(\epsilon^2)\right) du. \quad (3.51)$$

These integrals are elementary:

$$\int \frac{du}{\left(1 + u^2\right)^{\frac{1}{2}}} = \ln\left(u + \sqrt{1 + u^2}\right),$$

$$\int \frac{du}{\left(1 + u^2\right)^{\frac{3}{2}}} = \frac{u}{\sqrt{1 + u^2}},$$

leading to

$$T = \frac{r_0}{c} \left[u_B - u_A + \epsilon \left\{ \ln\left(\frac{u_B + \sqrt{1 + u_B^2}}{u_A + \sqrt{1 + u_A^2}}\right) - \frac{u_B}{2\sqrt{1 + u_B^2}} \right. \right.$$
$$\left. \left. + \frac{u_A}{2\sqrt{1 + u_A^2}} \right\} \right] + O(\epsilon^2).$$

The Shapiro time delay is the difference between the preceding T and the time that would be naïvely taken in the flat metric, $T_0 = \frac{x_A - x_B}{c} = \frac{r_0(u_B - u_A)}{c}$:

$$\Delta T = T - T_0$$
$$= \epsilon \frac{r_0}{c} \left\{ \ln\left(\frac{u_B + \sqrt{1 + u_B^2}}{u_A + \sqrt{1 + u_A^2}}\right) - \frac{u_B}{2\sqrt{1 + u_B^2}} + \frac{u_A}{2\sqrt{1 + u_A^2}} \right\} + O(\epsilon^2).$$
$$(3.52)$$

---

[7] The radius of the orbit of Venus is 1.08×10^{11} m, and the radius of the Earth's orbit is 1.50×10^{11} m.

If we ignore the terms of order ϵ^2 and higher, this formula should be correct to one part in a million; but a further simplification can be made by noting that, for the Earth and Venus, u_B and u_A in (3.50) are of the order of 10^2. In fact, u_A^2 and u_B^2 are both greater than 10,000 and we can approximate $1 + u^2$ with u^2 in Equation (3.52) and still get an answer that is correct to one part in ten thousand. We have

$$\frac{u_B}{\sqrt{1 + u_B^2}} = 1 + O\left(\frac{1}{u_B^2}\right),$$

$$u_B + \sqrt{1 + u_B^2} = 2u_B + O\left(\frac{1}{u_B}\right),$$

while, since u_A is negative,

$$\frac{u_A}{\sqrt{1 + u_A^2}} = \frac{-|u_A|}{\sqrt{1 + u_A^2}} = -1 + O\left(\frac{1}{u_A^2}\right),$$

$$u_A + \sqrt{1 + u_A^2} = |u_A|\left(-1 + \sqrt{1 + \frac{1}{u_A^2}}\right) = \frac{1}{2|u_A|} + O\left(\frac{1}{u_A^3}\right).$$

Substituting these expressions in (3.52) reduces it to

$$\Delta T = \frac{r_S}{c}\left(\ln\left(4u_B|u_A|\right) - 1\right) + \cdots, \tag{3.53}$$

where $r_S = \epsilon r_0 = \frac{2GM}{c^2}$. This expression should be valid to one part in 10^4 for the Earth and Venus. With the numerical values, (3.50) in (3.53) gives $\ln(4u_B|u_A|) - 1 = 10.8$ and

$$\Delta T = 107\,\mu s.$$

This can be tested by bouncing radar signals off Venus when it appears close to, but is on the opposite side of, the Sun. The signals must go there and back, and they return $214\,\mu s$ later than would naïvely be expected.

The Shapiro time delay also affects signals between the Earth and distant inter-planetary probes, such as Pioneer and Voyager, and this must be allowed for in all radio communications with such probes. For example, Saturn is 1.433×10^{12} m from the Sun, some 10 times further than the Earth; so, if a probe is out at Saturn's orbit, then B is Saturn ($u_B = 2,059$) and A is the Earth ($u_A = -216$) and the excess time taken to send a signal from the Earth to the probe when it is on the other side of the Sun is $132\,\mu s$. When the spacecraft and the Sun are at opposite points in the sky (i.e. the Earth and the spacecraft are both on the same side of the Sun and in line with it), Equation (3.52) can be used, except u_A and u_B are both positive, in which case it approximates to

$$\Delta T = \frac{r_S}{c} \ln\left(\frac{u_B}{u_A}\right) + \cdots . \qquad (3.54)$$

With this configuration, the Sun is not in the way and r_0 can be zero (when the Sun, the Earth, and the spacecraft are exactly in line), but r_0 cancels out in (3.54) and $\frac{u_B}{u_A} = \frac{x_B}{x_A}$ is finite. Again taking x_B to be the radius of Saturn's orbit, we get a time delay of 22 μs.

Lastly, we emphasise again that in our discussion here we have drawn the trajectory of the beam of light in Figure 3.3 as a straight line, but we have already seen in the last section that the gravitational field of the Sun will also cause the light beam to bend. We could modify the discussion to allow for this, but that would only add a correction of order ϵ^2 to Equation (3.52), which again is negligible.

3.7.4 The Event Horizon and Black Holes

As a star gets more and more compact and its size gets close to r_S, some very strange things happen. Consider a beam of light leaving the surface of a star, at $r_* \geq r_S$, and travelling radially outwards. As it emerges, the angles θ and ϕ are constant and $\mathcal{L}$ in (3.21) is zero, so

$$\frac{dr}{dt} = \left(1 - \frac{2GM}{rc^2}\right).$$

A stationary observer a long way away, at fixed $r = r_O \gg r_* \geq r_S$, has a clock that measures time $\tau_O = \sqrt{\left(1 - \frac{r_S}{r_O}\right)} t \approx t$. If light leaves the surface of the star at time $\tau_O = 0$ on the observer's clock, it will reach the observer at time

$$\tau_O = \frac{1}{c} \int_{r_*}^{r_O} \frac{dr}{1 - \frac{r_S}{r}} = \frac{r_O - r_*}{c} + \frac{r_S}{c} \ln\left(\frac{r_O - r_S}{r_* - r_S}\right). \qquad (3.55)$$

As $r_* \to r_S$ this diverges logarithmically: it takes an infinite amount of the observer's time for the light to reach them, so the observer will never see the light. (This is for finite r_O, all we have assumed is that $r_O \gg r_*$.)

Even weirder is if the star is so compact that its surface lies inside the Schwarzschild radius $r_* < r_S$ and the light starts from some point with $r < r_S$. Then the terms in the Schwarzschild line element (2.31) have a very different interpretation. Re-arranging the line element as

$$ds^2 = \left(\frac{2GM}{c^2 r} - 1\right) c^2 dt^2 - \frac{dr^2}{\left(\frac{2GM}{c^2 r} - 1\right)} + r^2\left(d\theta^2 + \sin^2\theta\, d\phi^2\right), \qquad (3.56)$$

we can see that fixing r, θ, and ϕ and allowing t to vary results in a space-like separation, $d^2 s > 0$, while fixing t, θ, and ϕ and allowing r to vary results in a time-like separation, $d^2s < 0$. The physical interpretation of this is that, for $r < r_S$, 't' is a space-like coordinate and 'r' is a time-like coordinate – it is as though space and time are interchanged. To emphasise this, change the notation and let $\tilde{r} = ct$ and $\tilde{t} = (r_S - r)/c$, with $0 < \tilde{t} < \frac{r_S}{c}$ in the range $r_S > r > 0$ while $-\infty < \tilde{r} < \infty$. Define

$$\tilde{t}_\Omega = \frac{r_S}{c} = \frac{2GM}{c^3}$$

so that $0 < \tilde{t} < \tilde{t}_\Omega$; then the line element is

$$ds^2 = -\left(\frac{\tilde{t}_\Omega}{\tilde{t}} - 1\right)c^2 d\tilde{t}^2 + \frac{\tilde{t}\, d\tilde{r}^{\,2}}{(\tilde{t}_\Omega - \tilde{t})} + c^2(\tilde{t}_\Omega - \tilde{t})^2\left(d\theta^2 + \sin^2\theta\, d\phi^2\right). \quad (3.57)$$

This is a weird geometry in which everything is highly time-dependent: spheres of constant $\tilde{r}$ have radius $c(\tilde{t}_\Omega - \tilde{t})$ and shrink to nothing as $\tilde{t}$ increases from 0 to $\tilde{t}_\Omega$; a ruler aligned radially at fixed θ and ϕ, with $d\tilde{r} \neq 0$, is stretched to infinite length as $\tilde{t}$ approaches $\tilde{t}_\Omega$. Three-dimensional space is highly anisotropic and collapses to a 1-dimensional line as $\tilde{t} \to \tilde{t}_\Omega$, where an observer would be crushed in two directions and indefinitely stretched in the third, and time stops.

At the end of the world, $\tilde{t} = \tilde{t}_\Omega$ (hence the subscript Ω), tidal forces become infinite, and there is a singularity in the geometry. (This is not just a bad choice of coordinates; there is no choice of coordinates that will remove this singularity.) There is no such thing as infinity in physics. If our equations give an infinite answer for a physical quantity, then the equations are wrong; the mathematical model has broken down. No one knows what really happens at $\tilde{t} = \tilde{t}_\Omega$ (equivalently $r = 0$); the mathematical structure of Einstein's general theory of relativity is simply not a correct description of the real world and must be replaced by a more sophisticated model, as yet unknown, possibly requiring a quantum theory of gravity. (See the discussion on page 166.)

We can visualise what is happening here by examining the structure of the light cones. Go back to ct–r coordinates and change the coordinates yet again by defining

$$r' = r + r_S \ln\left|\frac{r - r_S}{r_S}\right|.$$

(The modulus signs ensure that r' is well defined both for $r > r_S$ and $r < r_S$.) Then

$$dr' = \frac{dr}{1 - \frac{r_S}{r}}$$

and, for a light ray,

$$0 = ds^2 = \left(1 - \frac{r_S}{r(r')}\right)\left(-c^2 dt^2 + (dr')^2\right) + r^2(r')\left(d\theta^2 + \sin^2\theta d\phi^2\right).$$

Radially directed light rays have θ and ϕ constant so

$$dr' = \pm c\, dt.$$

For $r > r_S$, the plus sign is for rays moving radially outward and the minus sign for rays moving radially inward. In any case, a radially moving light ray is characterised by

$$v = ct + r' = constant$$

and, using (v, r, θ, ϕ) as coordinates[8]

$$ds^2 = -\left(1 - \frac{r_S}{r}\right)dv^2 + 2dvdr + r^2\left(d\theta^2 + \sin^2\theta d\phi^2\right), \tag{3.58}$$

which clearly vanishes when v, θ, and ϕ are constant, but it also vanishes along curves with θ and ϕ constant and

$$dv = \frac{dr}{\left(1 - \frac{r_S}{r}\right)}. \tag{3.59}$$

These are radially outward directed light rays for $r > r_S$ (r increases as v increases) and radially inward for $r < r_S$. We can visualise what is happening here by drawing light cones of (3.58) in the v-r plane. Constant v always corresponds to radially inward moving light rays, while solutions of

$$\frac{dv}{dr} = \frac{1}{1 - \frac{r_S}{r}} \quad \Rightarrow \quad v(r) = r + r_S \ln\left|\frac{r - r_S}{r_S}\right| + constant$$

represent radially outward moving light rays for $r > r_S$ and radially inward moving light rays for $r > r_S$.

A beam of light can pass from $r > r_S$ to $r < r_S$ but cannot pass from $r < r_S$ to $r > r_S$ – the sphere at $r = r_S$ acts like a one-way membrane for light. Nothing, not even light, can escape from the interior of the sphere at r_S, and such a sphere is called a *black hole*. No event occurring at a point $r < r_S$ can ever be seen by an observer at $r > r_S$, and the sphere $r = r_S$ is called the *event horizon* of the black hole. From (3.55) light from an event just outside the black hole, with $r > r_*$ but only just, takes an infinite amount of time to reach a distant observer. Like the world in *The Lion, the Witch and the Wardrobe* of *The Chronicles of Narnia*, the interior geometry is a very different world with a different

---

[8] These are known as *Eddington–Finkelstein* coordinates.

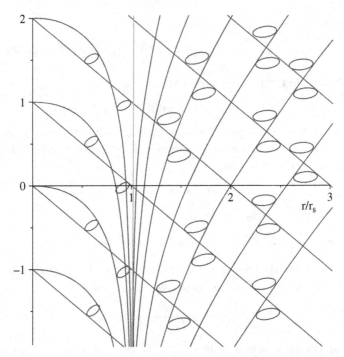

Figure 3.4 **Light cones in the Schwarzschild geometry.** The horizontal axis is $\frac{r}{r_S}$, so the event horizon is at $\frac{r}{r_S} = 1$. The straight lines running from the bottom right to the top left are lines of constant v, corresponding to inward going light rays. For $r > r_S$, there are both inward and outward going light rays; for $r < r_S$, the entire light cone is tipped over to point inward towards $r = 0$ ($\tilde{t} = \tilde{t}_\Omega$), and all light rays are inward going.

notion of time; but unlike Narnia, once you are through the back of the wardrobe, there is no getting back!

Although it is difficult to see black holes directly, because they cannot emit light, there is substantial evidence for their existence. Their gravitational influence can be detected, for example, and it transpires that some galaxies harbour large black holes at their centres, millions of times more massive than the Sun. Figure 3.6 shows the orbits of stars close to our own galactic centre; there is clearly a very massive compact object at the centre influencing them. It is a black hole with an estimated mass of 4 million times that of the Sun, which would have a Schwarzschild radius $r_S = 1.2 \times 10^7$ km, so the innermost stable circular orbit would have radius 3.6×10^7 km (see question 3.8). Hot gas swirling around the black hole was imaged directly in 2022 by the Event

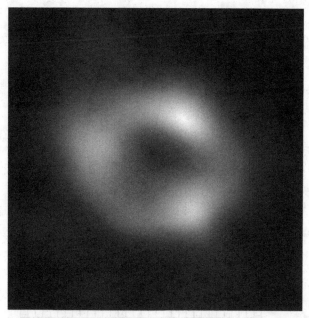

Figure 3.5 **Supermassive black hole at the centre of our Galaxy**. A composite image, taken using radio waves, of hot gas swirling around the black hole at the centre of our Galaxy.
©EHT collaboration,
https://eventhorizontelescope.org

Horizon Telescope (EHT; see Figure 3.5), using radio waves emitted by matter falling into the black hole, and found to have an angular diameter of 50 μas (1 micro-arcsecond is 4.8×10^{-12} radians). The centre of the Galaxy is 26,000 light years away, so the radius of the EHT image is 3×10^7 km.

Our neighbouring Andromeda galaxy is believed to have a supermassive black hole at its centre estimated to be 40 million solar masses, and there are even bigger monsters out there; the giant elliptical galaxy M87,[9] in the constellation of Virgo, is believed to harbour a black hole more than a billion times more massive than the Sun.

The tidal forces associated with such large black holes need not be extreme for an astronaut. At the event horizon of the 4 million solar mass black hole at the centre of our Galaxy, with a radius of 30×10^6 km, the tidal forces over 2 m are negligible. If it were not for the hot gasses

[9] Number 87 in Messier's catalogue of nebulae.

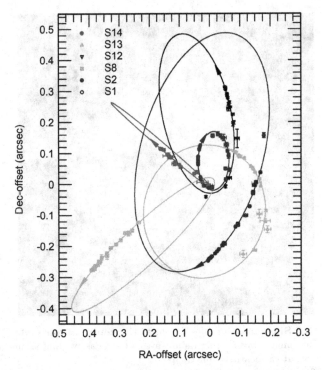

Figure 3.6 **Orbits of stars close to the Galactic centre.** The star S2 (the small vertical ellipse) has a 15-year orbit and is 1.8×10^{13} m from the central object at its closest. From the orbital dynamics of S2 the central object's mass is calculated to be 4.1×10^6 times the mass of the Sun. No known object can contain this amount of mass in such a small volume of space except a black hole.

Reproduced with permission from: *SINFONI in the Galactic Center: Young Stars and Infrared Flares in the Central Light-Month.* Eisenhauer, F. *et al.* *The Astrophysical Journal* **628** (2005), 246.

swirling around, an adventurous astronaut could sail in through the event horizon without necessarily noticing anything unusual for a while, until she got closer to the singularity at $r = 0$, or tried to turn round and come back!

More will be said on the subject of black holes in §6.2.

3.7.5 Gravitational Lensing

An interesting consequence of the light deflection discussed in §3.7.2 is that, if the imaged star is exactly in line with the centre of the Sun, then

Figure 3.7 **Gravitational lensing of one galaxy by another.** The light from a distant galaxy is warped into a ring by the gravity of a closer, massive galaxy which is the bright spot in the centre of the ring.
Image courtesy of NASA/European Space Agency/Hubble

there is no way of determining the angle about the x-axis in Figure 3.2. From symmetry, all such angles are equivalent, and the light of the star will be stretched in to a thin ring of angular diameter $2\phi_\infty$ centred on the Sun. Such a ring is called an *Einstein ring*.

In practice, if this did happen, the ring would be far too faint to be seen, even during an eclipse. However, for a more extended object, such as a galaxy rather than a star, it is quite possible that a direct line will pass through the galaxy, and the image of the whole galaxy will be distorted in a phenomenon called *gravitational lensing*. The galaxy will not be exactly circular, of course, and it is very unlikely that the x-axis will pass through the centre of the galaxy; the galaxy is more likely to be off-centre and appear to be an arc rather than a ring, as in Figure 3.7, in which the lensed and the lensing objects are both galaxies.

This can be observed if the deflecting mass, unlike the Sun, is not shining. For obvious reasons, such a mass is called *dark matter*. This phenomenon can be used to estimate the mass of the intervening object, even when it's not actually visible. More will be said about dark matter in Chapter 7.

3.8 Gravitational Waves

An oscillatory distortion of space-time is like a gravitational wave. A line element that represents such a wave travelling in the x-direction is[10]

$$ds^2 = -c^2 dt^2 + dx^2 + \left\{ 1 - P \cos \left(\omega \left(t - \frac{x}{c} \right) \right) \right\} dy^2$$
$$+ \left\{ 1 + P \cos \left(\omega \left(t - \frac{x}{c} \right) \right) \right\} dz^2. \tag{3.60}$$

For $P \ll 1$, this is a small perturbation of Minkowski space-time.

Gravitational waves should not be confused with gravity waves in a fluid. A gravitational wave propagates through a vacuum, empty space-time containing no matter, much like an electromagnetic wave can propagate through a vacuum. It is necessary that there be some oscillating matter somewhere in the Universe to produce a gravitational wave, just as there must be some oscillating electric charges in order to produce an electromagnetic wave; but once formed these waves can travel through a vacuum.[11] A gravity wave is a consequence of a fluid, such as water, oscillating in a gravitational field; gravity waves have been observed since time immemorial, by people standing on beaches. Gravitational waves are much harder to detect and were not observed directly until 2016, an effort involving a huge team of physicists, both theorists and experimentalists, and securing the Nobel Prize in Physics in 2017.

The Lagrangian determining the motion of a free particle following a geodesic in a space-time described by such an oscillating metric is

$$\mathcal{L}(t, x, y, z; \dot{t}, \dot{x}, \dot{y}, \dot{z}) = -c^2 \dot{t}^2 + \dot{x}^2 + \left\{ 1 - P \cos \left(\omega \left(t - \frac{x}{c} \right) \right) \right\} \dot{y}^2$$
$$+ \left\{ 1 + P \cos \left(\omega \left(t - \frac{x}{c} \right) \right) \right\} \dot{z}^2$$

and the equations of motion are

$$c^2 \ddot{t} + \frac{1}{2} P \omega \sin \left(\omega \left(t - \frac{x}{c} \right) \right) (-\dot{y}^2 + \dot{z}^2) = 0$$
$$\ddot{x} + \frac{1}{2c} P \omega \sin \left(\omega \left(t - \frac{x}{c} \right) \right) (-\dot{y}^2 + \dot{z}^2) = 0$$

---

[10] This is only an approximate solution of Einstein's equations for $P \ll 1$; see §6.2. The notation P stands for polarisation, as explained in that section.

[11] The production of an observable gravitational wave requires a cataclysmic event, and the mathematical analysis involves a lot of computing power; it is a much harder problem than understanding the production of an electromagnetic wave.

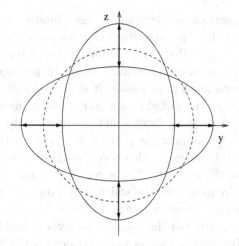

Figure 3.8 **A gravitational wave.** A circle in the y-z plane distorts into an oscillating ellipse as a gravitational wave moving in the x-direction passes through.

$$\frac{d[\{1 - P\cos\left(\omega\left(t - \frac{x}{c}\right)\right)\}\dot{y}]}{d\tau} = 0$$

$$\frac{d[\{1 + P\cos\left(\omega\left(t - \frac{x}{c}\right)\right)\}\dot{z}]}{d\tau} = 0.$$

The simplest solution, consistent with $\mathcal{L} = -c^2$, is $t = \tau$ with x, y, and z all constant. To visualise what this solution means, consider the $x = 0$ plane at a fixed time. At any time t the square of the distance d from the spatial origin to an arbitrary point in the y-z plane is

$$d^2 = \{1 - P\cos(\omega t)\}y^2 + \{1 + P\cos(\omega t)\}z^2,$$

and the locus of such points is an ellipse,

$$\frac{y^2}{a^2} + \frac{z^2}{b^2} = 1,$$

with semi-major axis $a = \dfrac{d}{\sqrt{1 - P\cos(\omega t)}}$ and semi-minor axis $b = \dfrac{d}{\sqrt{1 + P\cos(\omega t)}}$ (assuming $P\cos(\omega t) > 0$). The eccentricity is

$$\epsilon = \sqrt{1 - \frac{b^2}{a^2}} = \sqrt{1 - \frac{1 - P\cos(\omega t)}{1 + P\cos(\omega t)}} \approx \sqrt{2P\cos(\omega t)}$$

(more generally $\epsilon = \sqrt{2|P\cos(\omega t)|}$). This describes an oscillating ellipse in the y-z plane, as shown in Figure 3.8.

In electromagnetism, accelerating charges produce electromagnetic waves, and gravity has the same feature; accelerating masses can produce gravitational waves. But gravity is such a weak force that it requires extremely large masses, astronomically large, to produce a detectable wave. As a gravitational wave passes, it distorts lengths, as shown in Figure 3.8, and it was predicted that a pair of coalescing binary neutron stars 100 million light years away would produce a distortion in length, called the strain $\frac{\Delta L}{L}$, of only one part in 10^{21}. The Earth is 12,700 km $\approx 12 \times 10^6$ m in diameter, so this corresponds to measuring a length of $10^{-21} \times 1.2 \times 10^7$ m $\approx 10^{-14}$ m across the diameter of the Earth. This would require measuring the width of 10 protons across a distance of 13,000 kilometres – truly a formidable task. Yet a small number of physicists in the 1960s had the audacity to try and build a detector to measure this. Gravitational waves were first detected in 2016, 100 years after Einstein first predicted them, and it took more than 50 years to develop the technology to do it. The detector was not 13,000 kilometres in size; it was 4 km long, so the experimentalists succeeded in measuring a length change of only 10^{-18} metres, 1/1,000th the size of a proton.

Variations in length produce variations in the phase in a beam of light. Interferometers are very good at detecting tiny variations in phase, and the detection was made with a special laser interferometer. Because of the geometry of the distortions, shown in Figure 3.8, it is best to have a laser interferometer with two long arms at right angles to each other. The arrangement used in the Laser Interferometer Gravitational-Wave Observatory (LIGO) in the United States has two arms 4 km long, so they were trying to detect length changes of $4 \times 10^3 \times 10^{-21}$ m $= 4 \times 10^{-18}$ m. Actually there were two independent detectors about 2,500 km apart, one at Hanford, Washington, and one in Livingston, Louisiana. The detection signal of the 2016 event is shown in Figure 3.9.

The expected signal for any chosen initial pair of masses with given orbital characteristics can be calculated numerically from the field equations of general relativity. A set of templates is produced, and these are compared with the measured signal to find the best fit for the masses and the orbital characteristics producing the signal. The surprise was that this event did not come from the merging of two solar mass neutron stars in a nearby galaxy a few hundred million light years away, as expected, but from two black holes of 26 and 39 solar masses 1.3 billion light years away, coalescing to form a single black hole of 60 solar masses. The peak

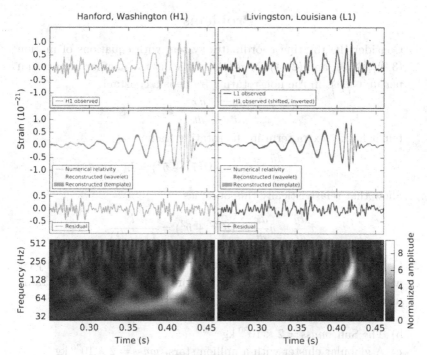

Figure 3.9 **The first gravitational wave ever detected directly.** Signals in two different detectors: the left-hand panels are Hanford; the right-hand panels are Livingston. In each case, the top figure is the detected signal of the strain $\frac{\Delta L}{L}$ as a function of time. The entire signal only lasts 0.2 s, and the peak strain is 10^{-21}. The shape of these signals allows some physical characteristics of the system to be determined, such as its distance, the individual masses of the two initial bodies at the start, and the final mass at the end after they have merged. The second panel from the top is the best-fit expected signal as calculated numerically from general relativity. The 'Frequency (Hz)' panel shows the measured frequency of the wave as a function of time. It is a feature of the general relativistic analysis of these events that the frequency increases sharply just before the merger, giving a characteristic 'chirp'.

Adapted from:

Observation of Gravitational Waves from a Binary Black Hole Merger, B. P. Abbott *et al.* (LIGO Scientific Collaboration and Virgo Collaboration). *Phys. Rev. Lett.* **116** (2016), 061102. https://doi.org/10.1103/PhysRevLett.116.061102

power output was about 50 times that of the whole visible Universe, though it only lasted for a fraction of a second. No one expected that. Gravitational waves produced by a pair of coalescing neutron stars were detected later, in 2017.

Problems

1) Consider the rotating coordinates system with equations of motion (3.9), (3.10), and (3.11). Non-relativistically the angular momentum per unit mass in the non-rotating frame is a constant,

$$x\frac{dy}{dt} - y\frac{dx}{dt} = l,$$

but it has an extra term in the rotating frame

$$\tilde{x}\frac{d\tilde{y}}{d\tilde{t}} - \tilde{y}\frac{d\tilde{x}}{d\tilde{t}} + \Omega(\tilde{x}^2 + \tilde{y}^2) = l.$$

Show that relativistically this becomes

$$\tilde{x}\frac{d\tilde{y}}{d\tilde{t}} - \tilde{y}\frac{d\tilde{x}}{d\tilde{t}} + \Omega(\tilde{x}^2 + \tilde{y}^2) = \frac{l}{\left(1 + \frac{l\Omega}{c^2}\right)}.$$

2) Calculate what the density of the following astronomical objects would be if they were compressed down to their Schwarzschild radii, using the naïve volume $4\pi r^3/3$:
   a) The Earth, $mass = 6 \times 10^{24}\,\mathrm{kg}$
   b) The Sun, $mass = 2 \times 10^{30}\,\mathrm{kg}$
   c) A globular cluster with a million stars, $mass = 2 \times 10^{36}\,\mathrm{kg}$
   d) The visible Universe, estimated mass $\approx 10^{53}\,\mathrm{kg}$.

3) A clock is dropped from rest at $r = r_0$ and falls freely in a radial direction ($\dot{\theta} = \dot{\phi} = 0$) towards a black hole, from a coordinate distance r_0 down to r_1. Show that

$$\left(\frac{dr}{d\tau}\right)^2 = c^2\left(k^2 - 1 + \frac{2GM}{rc^2}\right),$$

where τ is the proper time of the clock, M the mass of the black hole, and k the integration constant defined in the text. Hence show that $k^2 = 1 - \frac{2GM}{r_0 c^2}$ and

$$\frac{dr}{dt} = \frac{\dot{r}}{\dot{t}} = -\left(1 - \frac{2GM}{rc^2}\right)\frac{\sqrt{\frac{2GM}{r} - \frac{2GM}{r_0}}}{\sqrt{1 - \frac{2GM}{r_0 c^2}}}.$$

Derive the time taken to fall from rest at r_0 down to r_1, with $r_0 \gg r_1 > \frac{2GM}{c^2}$, as measured by the falling clock,

$$\tau_c = \frac{r_0^{3/2}}{\sqrt{2GM}}\int_{\frac{r_1}{r_0}}^{1}\frac{\sqrt{y}}{\sqrt{1-y}}dy \approx \frac{\pi r_0^{3/2}}{2\sqrt{2GM}},$$

where $y = \frac{r}{r_0}$. For an observer fixed at $r = r_0$, the proper time on their clock is $d\tau_0 = \sqrt{1 - \frac{2GM}{r_0 c^2}}\, dt$. Show that the time the stationary clock at r_0 records for the falling clock to fall from r_0 down to r_1 is

$$\tau_0 = \sqrt{1 - \frac{2GM}{r_0 c^2}} \int_0^t dt = r_0^{\frac{3}{2}} \frac{\left(1 - \frac{2GM}{r_0 c^2}\right)}{\sqrt{2GM}} \int_{\frac{r_1}{r_0}}^1 \frac{y^{\frac{3}{2}}\, dy}{(1-y)^{1/2}\left(y - \frac{r_s}{r_0}\right)}$$

$$\approx \frac{r_0^{\frac{3}{2}}}{\sqrt{2GM}} \int_\epsilon^1 \frac{y^{\frac{3}{2}}\, dy}{(1-y)^{1/2}\left(y - \epsilon\frac{r_s}{r_1}\right)},$$

where $\epsilon = \frac{r_1}{r_0}$ and again the approximation is for $r_0 \gg r_1 > r_s = \frac{2GM}{c^2}$. The second integral is infinite as $r_1 \to r_s$. What does this mean?

4) Show that the radial component of the geodesic equation in a Schwarzschild metric leads to the general relativistic orbit equation

$$\frac{d^2 u}{d\phi^2} + u = \frac{r_s^2 c^2}{l^2} + \frac{3}{2}u^2,$$

where $u = \frac{2GM}{rc^2}$.

5) The general relativistic orbit equation can be written

$$\dot{r}^2 - \frac{c^2 r_s}{r} + \frac{l^2}{r^2} - \frac{l^2 r_s}{r^3} + c^2(1 - k^2) = 0.$$

Interpret this as an energy equation and sketch the effective potential. Analyse the potential energy of the general relativistic orbit equation, for various values of k and l.

a) Find the extrema of the potential energy and categorise them as maxima and minima. Describe the possible orbits for both bound and unbound motion.

b) Find the circular orbits, categorise them as stable or unstable, and find the smallest possible circular orbit as the parameters k and l are varied. Calculate the speed $r\frac{d\phi}{d\tau}$ in the smallest stable circular orbit, where τ is the proper time.

c) Show that there are no stable circular orbits for $r < 3r_s$.

d) Calculate the closest possible point of approach (perihelion) for a stable orbit, not necessarily circular.

6) Define coordinates v and u in Schwarzschild space-time by $v = ct + r'$ and $u = ct - r'$, where

$$r' = \int \frac{dr}{1 - r_s/r} = r + r_s \ln\left|\frac{r - r_s}{r_s}\right|.$$

Show that in these coordinates the metric becomes

$$ds^2 = -\left(1 - \frac{r_S}{r(u,v)}\right) dv\, du + r^2(u,v)(d\theta^2 + \sin^2\theta d\phi^2),$$

where $r(u,v)$ is determined by

$$(v-u)/2 = r(u,v) + r_S \ln\left|\frac{r(u,v) - r_S}{r_S}\right|.$$

Show that if we further define $v' = \exp(v/2r_S)$ and $u' = -\exp(-u/2r_S)$, then the metric becomes

$$ds^2 = \mp F^2(u',v') dv'\, du' + r^2(u',v')(d\theta^2 + \sin^2\theta d\phi^2),$$

where $F^2 = \left(\frac{4r_S^3}{r}\right)\exp(-r/r_S)$ and $r(u',v')$ is defined by

$$v'u' = -\frac{(r - r_S)}{r_S}\exp(r/r_S).$$

Does the metric in these coordinates look singular at $r = r_S$?

4

The Geometry of Curved Spaces

In this chapter the mathematics necessary for describing higher-dimensional curved spaces is developed. This topic is unavoidably more abstract than those of the other chapters in the book, and some more technical aspects are relegated to the appendices. The main results necessary for understanding the remaining chapters are summarised on page 116. This is all that is really needed for Chapters 5, 6, and 7.

4.1 Inner Products

The Lagrangians that we were using in the previous section were all of the form

$$\mathcal{L} = g_{\mu\nu}\big(x(\tau)\big)\dot{x}^\mu(\tau)\dot{x}^\nu(\tau),$$

where

$$\dot{x}^\mu = \frac{dx^\mu}{d\tau}.$$

Now $\frac{dx^\mu}{d\tau}$ is a vector, the velocity vector (a 4-vector in space-time) tangent to the trajectory $x^\mu(\tau)$. If τ is the proper time and $x^\mu(\tau)$ is a geodesic, then the Lagrangian is the length squared of the 4-velocity, which is negative (or zero), since the 4-velocity is time-like (or light-like).

In general, the metric gives us a way of defining dot products for vectors. In 3-dimensional Euclidean space the dot product between two vectors $\overrightarrow{W}$ and $\overrightarrow{V}$ is often written (in Cartesian coordinates)

$$\overrightarrow{W}.\overrightarrow{V} = W^\alpha V^\alpha,$$

but, in the light of (2.3), this is better thought of as

$$\vec{W}.\vec{V} = \delta_{\alpha\beta} W^\alpha V^\beta,$$

which has a natural extension to a more general coordinate system:

$$\vec{W}.\vec{V} = g_{\alpha\beta} W^\alpha V^\beta.$$

In polar coordinates on a sphere of radius a, for example, $g_{\alpha\beta}$ $= a^2 \begin{pmatrix} 1 & 0 \\ 0 & \sin^2\theta \end{pmatrix}$, $\vec{W}$ has components (W^θ, W^ϕ), and $\vec{V}$ has components (V^θ, V^ϕ), hence

$$\vec{W}.\vec{V} = a^2(W^\theta V^\theta + \sin^2\theta\, W^\phi V^\phi).$$

In particular, putting $\vec{V} = \vec{W}$ gives the length squared of $\vec{W}$ on a sphere,

$$|\vec{W}|^2 = \vec{W}.\vec{W} = g_{\alpha\beta} W^\alpha W^\beta = a^2\{(W^\theta)^2 + \sin^2\theta(W^\phi)^2\}.$$

In fact, we need a metric to associate a length with a vector; a length is not defined without a metric. Mathematicians use the term *inner product* for this method of using the metric to combine two vectors to get a scalar. The dot product for vectors in Euclidean space is synonymous with an inner product.

A convenient notation is to associate a different kind of vector,

$$W_\alpha = g_{\alpha\beta} W^\beta,$$

with $\vec{W}$, so

$$W_\alpha = (W_\theta, W_\phi) = (a^2 W^\theta, a^2 \sin^2\theta\, W^\phi).$$

Then we can write

$$\vec{W}.\vec{W} = W_\alpha W^\alpha$$

or indeed

$$\vec{W}.\vec{V} = W_\alpha V^\alpha$$

for the dot product, or inner product, of any two vectors. W_α here are often called the components of the *covector*, associated with the vector $\vec{W}$ by the metric.

Vectors and covectors are completely different objects; it is only when a metric is defined that a specific covector can be associated with a given vector and vice versa. For this reason it is very important to keep track of the positions of the indices: a superscript tells us we are dealing with the components of a vector; a subscript represents the components of a covector. Do not get them mixed up.

This is a very important point which is sometimes hard to appreciate if you are new to the geometry of curved spaces: vectors and covectors are very different things. Usually, when you are introduced to Euclidean geometry, no distinction is made between them, because the metric in Cartesian coordinates looks like the identity matrix $\delta_{\alpha\beta}$, and the components of a covector look identical to those of the vector it is associated with; but even then, there is still a difference. For example, in 2-dimensional Euclidean space, if we rotate the axes through an angle θ the components of a vector change,

$$\begin{pmatrix} V^x \\ V^y \end{pmatrix} \;\to\; \begin{pmatrix} V^{x'} \\ V^{y'} \end{pmatrix} = R(\theta) \begin{pmatrix} V^x \\ V^y \end{pmatrix} = \begin{pmatrix} \cos\theta & \sin\theta \\ -\sin\theta & \cos\theta \end{pmatrix} \begin{pmatrix} V^x \\ V^y \end{pmatrix}.$$

Inner products, however, are invariant:

$$\vec{W}.\vec{V} = W_\alpha V^\alpha = W_{\alpha'} V^{\alpha'}.$$

This is usually understood by writing W_α as a 'row vector', so

$$\left(W_x, W_y \right) \;\to\; \left(W_{x'}, W_{y'} \right) = \left(W_x, W_y \right) R^T(\theta).$$

In index notation $V^\alpha \to V^{\alpha'} = R^{\alpha}{}_\beta V^\beta$ while $W_\alpha \to W_{\alpha'} = W_\beta \left(R^T \right)^{\beta}{}_{\alpha'}$ and

$$\vec{W}.\vec{V} = W_\alpha \left(R^T R \right)^{\alpha}{}_\beta V^\beta = W_\alpha V^\alpha$$

because $R^T = R^{-1}$ is an orthogonal matrix. A vector is rotated by R, while a covector is rotated by R^{-1}.[1]

In curved spaces, and indeed even in Euclidean space if curvilinear coordinates are being used, the distinction between vectors and covectors becomes more acute; they are very different animals.

In space-time the distinction between vectors and covectors is even more important. Even in flat space-time using Cartesian coordinates, Minkowski space-time, a covector,

$$V_\mu = \eta_{\mu\nu} V^\nu = (V_0, V_1, V_2, V_3) = (-V^0, V^1, V^2, V^3),$$

---

[1] Take careful note of the positions of the indices, though: V^β is rotated with $R^{\alpha'}{}_\beta$ and W_β by $(R^T)^\beta{}_{\alpha'}$ with $(R^T)^\beta{}_{\alpha'} = R_{\alpha'}{}^\beta = \delta_{\alpha'\gamma'} R^{\gamma'}{}_\delta \delta^{\delta\beta}$, where β has been 'raised' using the Euclidean metric $\delta^{\delta\beta}$ while α' has been 'lowered' using $\delta_{\alpha'\gamma'}$. In contrast, $R^{-1}R = \mathbf{1}$, or $(R^{-1})^\alpha{}_{\gamma'} R^{\gamma'}{}_\beta = \delta^\alpha{}_\beta$, has nothing to do with the Euclidean metric; $\delta^\alpha{}_\beta$ here is just the identity matrix, not the Euclidean metric. The orthogonality condition, $R^T = R^{-1}$, on the other hand, is crucially dependent on the Euclidean metric.

has different components from the corresponding vector. The 4-dimensional inner product is then

$$\vec{V}.\vec{V} = \eta_{\mu\nu}V^\mu V^\nu = V^\mu V_\mu = -(V^0)^2 + (V^1)^2 + (V^2)^2 + (V^3)^2.$$

For example, we know that for a particle with rest mass m and 4-momentum $\vec{P}$ the momentum can be decomposed into time-like and space-like parts,

$$P^0 = \gamma(v)mc, \qquad \mathbf{P} = \gamma(v)m\mathbf{v},$$

where $\mathbf{v}$ is the particle's 3-velocity and $P^0 = E/c$. (E is the relativistic energy, $\gamma(v)mc^2$ for a particle with rest mass m moving with speed v.) The inner product,

$$\vec{P}.\vec{P} = -(P^0)^2+(P^1)^2+(P^2)^2+(P^3)^2 = \gamma^2(v)m^2(-c^2+\mathbf{v}.\ \mathbf{v}) = -m^2c^2,$$

is Lorentz invariant.

In the same way, the relativistic 4-velocity, $\vec{U}$, decomposes as

$$U^0 = \gamma(v)c, \qquad \mathbf{U} = \gamma(v)\mathbf{v}$$

and

$$\vec{U}.\vec{U} = \gamma^2(v)(-c^2 + v^2) = -c^2. \tag{4.1}$$

The length squared of the 4-velocity is constant; even a particle at rest has $U^0 = c$ and $\vec{U}.\vec{U} = -c$. This is why we found $\mathcal{L} = -c^2$ in the relativistic examples of geodesic motion in Chapter 3 when the Lagrangian is evaluated on trajectories parameterised by the proper time τ; in that case the Lagrangian is just $\mathcal{L} = \vec{U}.\vec{U} = -c^2$.

4.2 General Coordinate Transformations

One is always free to choose whatever coordinate system one wants to describe any given geometry, and we can change from one coordinate system x^μ to another $x^{\mu'} = (x')^\mu$ if we have explicit expressions for $\left(x'(x)\right)^\mu$ as functions of x^μ. Let ds^2 be the line element at some fixed point P of a curved space or space-time. We are free to use either x-coordinates and label P by $x|_P$, or to use x' coordinates and label P by $x'|_P$. We can change between these two coordinate systems using the chain rule,[2]

$$ds^2 = g_{\mu'\nu'}(x'|_P)dx^{\mu'}dx^{\nu'} = g_{\mu'\nu'}(x'|_P)\left(\frac{\partial x^{\mu'}}{\partial x^{\rho}}dx^{\rho}\right)\left(\frac{\partial x^{\nu'}}{\partial x^{\lambda}}dx^{\lambda}\right)$$
$$= g_{\rho\lambda}(x|_P)dx^{\rho}dx^{\lambda},$$

where primed indices μ', ν' refer to primed coordinates $x^{\mu'} = (x')^{\mu}$, $x^{\nu'} = (x')^{\nu}$, and so on, and the summation convention has been used. The components of the metric at P in the primed coordinate system are thus related to the components in the unprimed system by

$$g_{\rho\lambda}(x|_P) = \frac{\partial x^{\mu'}}{\partial x^{\rho}}\frac{\partial x^{\nu'}}{\partial x^{\lambda}}g_{\mu'\nu'}(x'|_P). \tag{4.2}$$

Physical quantities labelled with coordinates whose components transform like (4.2) under coordinate changes are called *tensors*. The metric is a rank-2 tensor; more generally, a rank-n tensor **T** has components which change as

$$T_{\mu_1\cdots\mu_n}(x|_P) = \frac{\partial x^{\nu'_1}}{\partial x^{\mu_1}}\cdots\frac{\partial x^{\nu'_n}}{\partial x^{\mu_n}}T_{\nu'_1\cdots\nu'_n}(x'|_P). \tag{4.3}$$

A covector is a rank-1 tensor.

To understand this in more detail, step back a minute and just consider functions. Suppose we have a function which depends on position in space $f(x)$, such as the temperature at different places, for example. If we move from the point labelled by x to a neighbouring point $x + \delta x$, the value of function will change and we can perform a Taylor expansion:

$$f(x) \to f(x) + \delta x^{\mu}\partial_{\mu}f(x) + O(\delta x^2).$$

The partial derivatives, $\partial_{\mu}f = \frac{\partial f}{\partial x^{\mu}}$, are the components of the gradient of the function. Under a change of coordinates,

$$\partial_{\mu}f \to \partial_{\mu'}f = \frac{\partial x^{\nu}}{\partial x^{\mu'}}\partial_{\nu}f,$$

and we see that the gradient is a covector.

In writing the laws of physics in a general geometry, we shall demand that we have the freedom to use any coordinate system we wish and the equations of physics should not change form if a different coordinate system is chosen; everything must be written in terms of tensors.

---

[2] In this section we shall generically use Greek indices near the middle of the alphabet to label coordinates, regardless of whether we are in space or space-time. The index μ here would be 0, 1, 2, or 3 if we are in a 4-dimensional space-time or 1, 2, or 3 if we are in a 3-dimensional space. Everything in this section is general.

There is an alternative view of the transformation $x \to x'(x)$ which is sometimes more useful. Think of a fluid flow on the surface of a sphere, like the wind on the Earth's surface. At each point we have a velocity vector $\vec{V}$ which we can think of as a tangent vector to the flow lines of the fluid. In a short time interval τ, a fluid particle at x moves to a different point x' where

$$x^\mu \to x^{\mu'} = x^\mu + \tau V^\mu + O(\tau^2), \tag{4.4}$$

so

$$\frac{\partial x^{\mu'}}{\partial x^\nu} = \delta^\mu{}_\nu + \tau \frac{\partial V^\mu}{\partial x^\nu} + O(\tau^2) \quad \text{or} \quad \frac{\partial x^\nu}{\partial x^{\mu'}} = \delta^\nu{}_\mu - \tau \frac{\partial V^\nu}{\partial x^\mu} + O(\tau^2). \tag{4.5}$$

Suppose the temperature $T(x)$ of the fluid is non-uniform and varies from place to place. We can Taylor expand the temperature,

$$T(x') = T(x) + \tau V^\mu \partial_\mu T(x) + O(\tau^2), \tag{4.6}$$

where again $\partial_\mu T$ is a shorthand notation for the covector components $\frac{\partial T}{\partial x^\mu}$. In this expansion $V^\mu \partial_\mu T$ is the value of the gradient of the temperature in the direction of the vector $\vec{V}$. Notice that the definition of the gradient itself, $\nabla_\mu T = \partial_\mu T$, is strictly speaking not a vector field; rather, it defines a covector field.

The metric components also transform:

$$g_{\mu\nu}(x') = g_{\mu\nu}(x) + \tau V^\rho \partial_\rho g_{\mu\nu}(x) + O(\tau^2), \tag{4.7}$$

and combining this with (4.5) gives

$$g_{\rho'\sigma'}(x') \frac{\partial x^{\rho'}}{\partial x^\mu} \frac{\partial x^{\sigma'}}{\partial x^\nu} = g_{\mu\nu}(x') + \tau \left(g_{\rho\nu}(x) \frac{\partial V^\rho}{\partial x^\mu} + g_{\mu\sigma}(x) \frac{\partial V^\sigma}{\partial x^\nu} \right) + O(\tau^2)$$

$$= g_{\mu\nu}(x) + \tau \{ V^\rho \partial_\rho g_{\mu\nu}(x) + g_{\rho\nu}(x) \partial_\mu V^\rho$$
$$+ g_{\mu\rho}(x) \partial_\nu V^\rho \} + O(\tau^2)$$

$$= g_{\mu\nu}(x) + \tau \mathcal{L}_{\vec{V}} g_{\mu\nu}(x) + O(\tau^2),$$

where

$$\mathcal{L}_{\vec{V}} g_{\mu\nu}(x) = V^\rho \partial_\rho g_{\mu\nu}(x) + g_{\rho\nu}(x) \partial_\mu V^\rho + g_{\mu\rho}(x) \partial_\nu V^\rho. \tag{4.8}$$

$\mathcal{L}_{\vec{V}} g_{\mu\nu}(x)$ is the derivative of $g_{\mu\nu}(x)$ in the direction of $\vec{V}$ and is called the *Lie derivative*[3] of $g_{\mu\nu}(x)$ with respect to the vector field $\vec{V}(x)$. (We shall see later that there are other kinds of derivative in a curved space.)

---

[3] Pronounced *Lee* – Sophus Lie was a Norwegian mathematician active in the second half of the nineteenth century.

Note that[4]

$$ds'^2 = g_{\mu'\nu'}(x')dx^{\mu'}dx^{\nu'}$$
$$= \left(g_{\mu\nu}(x) + \tau V^\rho \partial_\rho g_{\mu\nu}(x)\right)dx^\mu dx^\nu$$
$$+ \tau\left(\partial_\mu V^\rho g_{\rho\nu}(x) + (\partial_\nu V^\rho g_{\mu\rho}(x))\right)dx^\mu dx^\nu + O(\tau^2)$$
$$= \left(g_{\mu\nu}(x) + \tau \mathcal{L}_{\vec{V}} g_{\mu\nu}(x)\right)dx^\mu dx^\nu + O(\tau^2)$$

or

$$ds'^2 = ds^2 + \tau\left(\mathcal{L}_{\vec{V}} g_{\mu\nu}(x)\right)dx^\mu dx^\nu + O(\tau^2),$$

so the line element is not in general invariant under such a transformation.

This notion of using $x^{\mu'}(x)$ to move points around is called a *diffeomorphism*,[5] and (4.8) represents how the metric transforms under such a deformation. Moving things around by performing diffeomorphisms like this is also called *Lie transport*. The difference between a coordinate change (4.2) and a diffeomorphism (4.7) is somewhat analogous to the difference between the Eulerian and the Lagrangian approach to classical mechanics. We might call (4.2) an Eulerian change in coordinates and (4.7) a Lagrangian change in coordinates,[6] but diffeomorphism is a better word for the latter; it represents a kind of change in shape – the geometry morphs in a smooth (differentiable) way.

If it so happens that $\mathcal{L}_{\vec{V}} g_{\mu\nu} = 0$, then

$$g_{\mu\nu}(x) = g_{\mu'\nu'}(x') \tag{4.9}$$

and we have a *symmetry* associated with the deformation generated by $\vec{V}$. A simple example of this is shifting the origin of Cartesian coordinates in Euclidean space: $x^\mu \to x^\mu + x_0^\mu$ with constants x_0^μ leaves (2.2) invariant. Rotations about the origin also leave it invariant. Again, Lorentz transformations leave the Minkowski space-time line element (2.16) invariant. Another example is a constant shift in ϕ for the metric on a sphere (2.10): let $\phi' = \phi + \phi_0$ with ϕ_0 a constant in (2.10). Then $d\phi' = d\phi$, and the metric has exactly the same form in both the primed

---

[4] We can very rapidly run out of letters for the indices, so it is not uncommon to use the same set of letters for both the primed and the unprimed coordinates – they are distinguished only by the presence or absence of a prime.

[5] Since we are Taylor expanding, we are assuming that the coordinate transformation is analytic, and hence differentiable. A *morphism* in mathematics is a map that preserves some kind of structure.

[6] An alternative terminology is to call (4.2) a *passive* coordinate transformation and (4.7) an *active* coordinate transformation, but these terms seem to have fallen into disuse somewhat in the physics community.

and the unprimed coordinate systems; this is a consequence of the fact that the sphere is symmetric under rotations about the axis $\theta = 0$ (or $\theta = \pi$). In terms of Lie derivatives, shifting ϕ by a constant corresponds to a diffeomorphism using the vector $\vec{V}$ with $(V^\theta, V^\phi) = (0, 1)$, and it is easily checked that $\mathcal{L}_{\vec{V}} g_{\alpha\beta} = 0$, since $\partial_\phi g_{\alpha\beta} = 0$ and the components of $\vec{V}$ are constant.

4.3 Vectors

Defining a vector in a curved space is actually rather subtle. In our analogy with fluid flow and Lagrangian coordinates, a flow line has a tangent vector at each point, and so a vector can be associated with a first-order differential operator, the gradient

$$\vec{V} \quad \leftrightarrow \quad V^\mu \frac{\partial}{\partial x^\mu}. \tag{4.10}$$

In the geometry of curved space a vector field is not a passive object; it pushes things around, and we can even define an 'action' of a vector field on a function. For the case of the sphere, for example, the vector with components $V^\alpha = (0, 1)$, in polar coordinates (θ, ϕ) (with $\alpha = 1, 2$ and $x^1 = \theta$, $x^2 = \phi$), is $\vec{V} = \frac{\partial}{\partial \phi}$, and we say that this *generates* rotations about the $\theta = 0$ axis.

If this seems a little abstract, think again of the Taylor expansion of a function, such as the temperature (4.6). Write this as

$$T(x') = T(x) + \tau \mathcal{L}_{\vec{V}} T(x) + O(\tau^2),$$

where the Lie derivative of a function is defined to be

$$\mathcal{L}_{\vec{V}} T(x) = V^\mu \partial_\mu T(x) = \vec{V} T(x). \tag{4.11}$$

In a sense, the Lie derivative acting on functions *is* a vector field, $\mathcal{L}_{\vec{V}} = \vec{V}$.

It is often convenient to expand vectors in some basis

$$\vec{V} = V^\mu \vec{b}_\mu;$$

for diffeomorphisms, for example, we identify $\vec{b}_\mu$ with the derivative in the x^μ direction,

$$\vec{b}_\mu = \frac{\partial}{\partial x^\mu}. \tag{4.12}$$

Such a basis is called a *coordinate basis*, as it is defined in terms of a specific choice of coordinates. In a very real sense, Equation (4.10) can be written as an equality

$$\vec{V} = V^\mu \frac{\partial}{\partial x^\mu}. \tag{4.13}$$

We must move from the kindergarten notion that a vector is an arrow to the more adult concept that a vector is a differential operator.

It cannot be emphasised strongly enough that a coordinate basis is *not* in general orthonormal. If $\vec{V} = V^\mu \vec{b}_\mu$ and $\vec{W} = W^\nu \vec{b}_\nu$ then, since

$$\vec{V}.\vec{W} = \vec{b}_\mu.\vec{b}_\nu V^\mu W^\nu = g_{\mu\nu} V^\mu W^\nu$$

for all vectors, we must interpret $\vec{b}_\mu.\vec{b}_\nu$ as

$$\vec{b}_\mu.\vec{b}_\nu = g_{\mu\nu}(x); \tag{4.14}$$

the inner product of two vectors is not the multiplication of two vectors; it is a number that depends on position.

A set of coordinate basis vectors is only orthonormal if $g_{\mu\nu}(x) = \delta_{\mu\nu}$ (or $\eta_{\mu\nu}$ in space-time), and this cannot be true at all points unless we are using Cartesian coordinates in flat space. We are, however, free to choose whatever basis we wish and we can always use an orthonormal basis. For a sphere of radius a, for example, with metric (2.10),

$$\vec{b}_\theta.\vec{b}_\theta = a^2, \quad \vec{b}_\phi.\vec{b}_\phi = a^2 \sin^2\theta, \quad \text{and} \quad \vec{b}_\theta.\vec{b}_\phi = 0,$$

this coordinate basis is orthogonal, but it is not normalised to unity.[7] We can choose an orthonormal basis

$$\vec{b}_1 = \frac{1}{a}\vec{b}_\theta = \frac{1}{a}\frac{\partial}{\partial\theta}, \quad \vec{b}_2 = \frac{1}{a\sin\theta}\vec{b}_\phi = \frac{1}{a\sin\theta}\frac{\partial}{\partial\phi}, \tag{4.15}$$

which is defined so that[8]

$$\vec{b}_a.\vec{b}_b = \delta_{ab},$$

where $a, b = 1, 2$. (All derivatives are inactive in dot products: the ∂_θ in $\vec{b}_1$ does not differentiate the $\frac{1}{\sin\theta}$ in $\vec{b}_2$ in the dot product.) This does not mean the sphere is flat – the components of any metric will always be

---

[7] In fact $\vec{b}_\phi$ has zero length at the north and south poles; it vanishes at these points. This is because polar coordinates are not good coordinates at the north and south poles; this is another example of a coordinate singularity at a point where the geometry is perfectly regular.

[8] To emphasise the difference between a coordinate basis and an orthonormal basis it is useful to use indices $\mu, \nu, \rho, \ldots$ (or $\alpha, \beta, \gamma, \ldots$) for the former and $a, b, c, \ldots$ for the latter.

δ_{ab} (or η_{ab} in space-time) in an orthonormal basis. Only in flat space can a coordinate basis be an orthonormal basis everywhere (and even then, only if we use Cartesian coordinates). Many vector formulae look very different in coordinate and in orthonormal bases and it is very important to bear the distinction in mind.

A technical point is that, in a coordinate basis, the components of the covector $\vec{b}_\mu f(x) = \partial_\mu f(x)$ represent partial differentiation of a function $f(x)$ with respect to the coordinate x^μ; but the components of the same covector in an orthonormal basis represent the normalised gradient of $f(x)$ in the direction of $\vec{b}_a$, which is not, in general, a simple partial derivative. A consequence of this is that, while partial derivatives commute, $\partial_\mu\partial_\nu = \partial_\nu\partial_\mu$, in general repeated actions of $\vec{b}_a$ in an orthonormal basis do not. In (4.15), for example,

$$\vec{b}_1 = \frac{1}{a}\partial_\theta, \quad \vec{b}_2 = \frac{1}{a\sin\theta}\partial_\phi \quad \text{giving} \quad [\vec{b}_1, \vec{b}_2] = -\frac{1}{a}\cot\theta\,\partial_2.$$

4.4 Covariant Derivatives

The foundations of Euclidean geometry were laid down around 300 BC by the Greek mathematician Euclid. Euclid based his logic on five postulates, or axioms, which form the foundations of planar geometry:

1. It is possible to draw a straight line between any two points.
2. It is possible to extend any finite straight line segment indefinitely in either direction as a straight line.
3. It is possible to draw a circle centred on any point with any radius.
4. All right angles are equal to one another.
5. If three straight lines are drawn and one crosses the other two in such a way that the interior angles made with the other two sum to less than 180°, then the other two straight lines, if extended indefinitely, will meet on that side on which the angles sum to less than 180°.

The fifth postulate seems rather cumbersome (if you find the statement hard to follow, take a look at Figure 4.1). The fifth postulate is certainly not as elegant as the others, and it was long felt that it was somehow redundant and ought to be derivable from the other four. Many clever people sought a way to do this (including the twelfth-century Persian poet Omar Kayyám). The search was finally abandoned in the first half

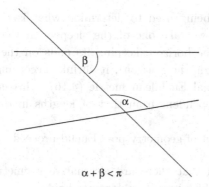

Figure 4.1 **Euclid's fifth postulate.** If $\alpha + \beta < \pi$, then two more horizontal lines, if extended to the right, will eventually meet.

of the nineteenth century, when it was discovered that there is a consistent geometry which satisfies the first four postulates but which violates the fifth, so it is not possible to derive the fifth postulate from the first four.[9] This new geometry is a space in which parallel lines appear to diverge when extended infinitely in either direction, due to the curvature of the space; it is a curved space with a constant negative curvature, the hyperbolic plane that we met in §2.5 (this is also sometimes called the Lobachevsky plane and will make another appearance later in Chapter 7 on Cosmology). This new geometry still requires a fifth postulate, but the postulate is not the same as that of Euclid.

Parallel lines appear to diverge in Lobachevsky's geometry because the metric is not the flat metric and space is curved. As the concepts of geometry developed after Lobachevsky's discovery it eventually emerged that the notion of 'parallel' can in fact be completely divorced from the notion of distance and length.

To illustrate this point, consider again Figure 2.1. We can use the flat 2-dimensional metric on this space, with line element written in 2-dimensional polar coordinates with the origin chosen at the vanishing point,

$$ds^2 = dr^2 + r^2 d\phi^2. \tag{4.16}$$

However, instead of using the 2-dimensional Euclidean notion of parallel, we shall use the railway tracks and fence wires to define parallel lines in the picture: we shall simply *decree* that the railway tracks and the fence wires are parallel. Then the 2-dimensional Euclidean metric on

---

[9] The surface of a sphere is obviously a curved space that does not satisfy the Euclidean postulates, but the second postulate makes it clear that we are only interested here in spaces for which lines can be extended indefinitely.

the plane is not being used to determine what is parallel and what is not. Indeed, if we take one of the sleepers in the foreground and slide it towards the horizon, keeping its ends on the tracks (parallel transport it towards the horizon), it shrinks according to the lengths of the 2-dimensional Euclidean metric (4.16) – this notion of parallel is completely independent from that of lengths in this 2-dimensional geometry.[10]

The development of geometry post-Euclid proceeded in two stages:

1. The realisation that there are alternative geometries (metrics) in which Euclid's fifth postulate does not hold.
2. The realisation that the definition of parallel can be completely divorced from the definition of lengths.

We need to develop mathematical machinery to handle these subtleties, so we need to think a little more carefully about how we define parallel vectors in a curved space.[11] This will lead us to a new way of differentiating vector fields, and other tensor fields, called *covariant differentiation*. We have already met a derivative; the Lie derivative of the metric was defined in (4.8). Lie derivatives are very important in studying symmetries, but they are only defined relative to a chosen vector field, $\vec{V}$ in Equation (4.8); they have no meaning without a vector field. We shall need another kind of derivative that does not require choosing a vector field. For example, the gradient of a function,

$$\nabla_\mu f = \partial_\mu f,$$

does not require introducing a vector field $\vec{V}$ but the Lie derivative,

$$\mathcal{L}_{\vec{V}} f = V^\mu \partial_\mu f,$$

does.

4.4.1 Connections

To express the laws of physics in a curved space-time, we shall need a generalisation of the gradient for vectors and higher-rank tensor fields.

---

[10] This notion of parallel is of course compatible with the 3-dimensional metric projected onto the 2-dimensional plane. The theory of perspective and projection was developed by the Italian painter Filippo Brunelleschi in 1413 and is a fascinating story in itself, but it is not directly relevant to general relativity.

[11] In this chapter we shall, for brevity, refer to 'space', but everything we say is equally applicable to a 4-dimensional space-time.

The simplest thing might be to just differentiate the components of a vector using partial derivatives with respect to the coordinates and try to define a derivative as

$$\frac{\partial W^\mu}{\partial x^\nu},$$

but this is not compatible with the way things change under coordinate transformations. For example, under $x^\mu \to x^{\mu'}(x)$ of the Eulerian kind in (4.2),

$$W^\mu(x|_P) \quad \to \quad W^{\mu'}(x|_P) = W^\nu(x'|_P)\frac{\partial x^{\mu'}}{\partial x^\nu},$$

we would have

$$\frac{\partial W^\mu}{\partial x^\nu} \quad \to \quad \frac{\partial W^{\mu'}}{\partial x^{\nu'}} = \frac{\partial x^\rho}{\partial x^{\nu'}}\frac{\partial}{\partial x^\rho}\left(\frac{\partial x^{\mu'}}{\partial x^\lambda}W^\lambda\right)$$

$$= \frac{\partial x^\rho}{\partial x^{\nu'}}\frac{\partial x^{\mu'}}{\partial x^\lambda}\frac{\partial W^\lambda}{\partial x^\rho} + W^\lambda\frac{\partial^2 x^{\mu'}}{\partial x^\lambda \partial x_\rho}\frac{\partial x^\rho}{\partial x^{\nu'}},$$

and the presence of the second term on the right-hand side means that $\frac{\partial W^\mu}{\partial x^\nu}$ is not a tensor (see Appendix A).

For covectors

$$\frac{\partial W_\mu}{\partial x^\nu} \quad \to \quad \frac{\partial W_{\mu'}}{\partial x^{\nu'}} = \frac{\partial x^\rho}{\partial x^{\nu'}}\frac{\partial W_{\mu'}}{\partial x^\rho} = \frac{\partial W_\lambda}{\partial x^\rho}\frac{\partial x^\lambda}{\partial x^{\mu'}}\frac{\partial x^\rho}{\partial x^{\nu'}} + W_\rho\frac{\partial^2 x^\rho}{\partial x^{\mu'}\partial x^{\nu'}},$$

and we can be clever and anti-symmetrise in μ' and ν', so

$$\partial_{\mu'}W_{\nu'} - \partial_{\nu'}W_{\mu'} = (\partial_\rho W_\lambda - \partial_\lambda W_\rho)\frac{\partial x^\rho}{\partial x^{\mu'}}\frac{\partial x^\lambda}{\partial x^{\nu'}},$$

which *is* a tensor. In fact, Maxwell's equations can be written purely in terms of such anti-symmetric derivatives (called *exterior derivatives*) and there is absolutely no need to introduce a fancy covariant derivative to write Maxwell's equations in a curved space-time. But Maxwell's equations are rather special in this regard, and covariant derivatives are necessary in more general situations.[12] The difficulty lies with the non-linearity of a general coordinate transformation, and such non-linearities

---

[12] In the relativistic formulation of Maxwell's equations the electric and magnetic fields E_α and B_α are combined into an anti-symmetric matrix,

$$F_{\mu\nu} = -F_{\nu\mu} = \begin{pmatrix} 0 & -E_x/c & -E_y/c & -E_z/c \\ E_x/c & 0 & B_z & -B_y \\ E_y/c & -B_z & 0 & B_x \\ E_z/c & B_y & -B_x & 0 \end{pmatrix}$$

which is an anti-symmetric rank 2 covariant tensor.

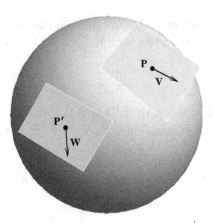

Figure 4.2 **Tangent planes to a sphere.** A vector $\overrightarrow{V}$ at a point P on a sphere lies in a plane tangent to the sphere at the point P. A vector $\overrightarrow{W}$ at a different point P' lies in a different plane, tangent to the sphere at the point P'.

are an essential feature of curved spaces. It is important to realise that, in a curved space, vectors do not really live in the space itself but are better thought of as living in a space that is tangent to the curved space at a point. For a 2-dimensional curved space we would call this a tangent plane (more generally a tangent space in three dimensions, or a tangent space-time if we are dealing with relativistic physics in four space-time dimensions). Every point has a different 2-dimensional tangent plane (Figure 4.2).

To differentiate a vector field $\overrightarrow{V}(x)$ we would want to take two points P and P' very close to one another, a distance $|x_{P'} - x_P|$ apart, and define something like

$$\lim_{P' \to P} \frac{(\overrightarrow{V}|_{P'} - \overrightarrow{V}|_P)}{|x_{P'} - x_P|}.$$

But $\overrightarrow{V}|_{P'}$ and $\overrightarrow{V}|_P$ live in *different* vector spaces, so how can we subtract them? This point cannot be emphasised strongly enough, so let's stress again, with feeling:

> Vectors living in different vector spaces
> cannot be added or subtracted!

We need a rule that relates vectors in the tangent plane at P to the tangent plane at P', a rule that connects vectors in the two tangent planes. We could choose some path between P and P' and Lie transport vectors from P to P', that results in the Lie derivative introduced on page 90, but the result would depend on the tangent vector to the path chosen. Instead, we shall give a rule that does not depend on any particular path or vector, but rather is intrinsic to the geometry of the space (though we can still choose a path later if we wish and use our rule to transport vectors along the path). This new rule, described in what follows, is called *parallel transport* and requires describing the notion of a *connection*.[13] It turns out that there is no unique rule; there are many possible different rules and many possible definitions of parallel, even for the same metric.

We may wish to impose some reasonable conditions on the rule; for example, we might demand that a vector does not change length when transported from P to P'. (Of course, this assumes that we have a metric.) A connection with this condition is called *metric compatible*, the connection defined by the rail tracks in Figure 2.1, for example, is not compatible with the 2-dimensional Euclidean metric (4.16). We shall only consider metric-compatible connections.

If we have a basis for vectors at P, $\vec{b}_\mu$ say, and a basis $\vec{b}'_\mu$ at P', then we only need to specify a transport rule for each of the basis vectors, and that will be sufficient to get the rule for any vector. Suppose we transport a basis vector $\vec{b}_\mu$ for the tangent space at P to the tangent space at P' and get a vector $\vec{c}'_\mu$ there. Then $\vec{c}'_\mu$ can be expanded in the basis $\vec{b}'_{\mu'}$ at P',

$$\vec{b}_\mu \quad \to \quad \vec{c}'_\mu = c^\nu{}_\mu \vec{b}'_\nu,$$

for some numerical coefficients $c^\nu{}_\mu$. If we use orthonormal bases $\vec{b}_a$ and $\vec{b}'_b$ at the two points (and the connection is metric compatible, so lengths don't change), then the corresponding matrix $c^b{}_a$ is just a rotation (or a Lorentz transformation for space-time), but it might not be so for a more general choice of bases, such as a coordinate basis.

In $P = P'$ we would have $\vec{b}_\nu = \vec{b}'_\nu$ and get $c^\nu{}_\mu = \delta^\nu{}_\mu$. If P' is infinitesimally close to P, say the separation is $|x_{P'} - x_P| = \epsilon$ with $\epsilon \ll 1$, we shall assume that

$$c^\nu{}_\mu(\epsilon) = \delta^\nu{}_\mu + \Gamma^\nu{}_\mu \epsilon + O(\epsilon^2),$$

---

[13] The term *parallel transport* is used in analogy with Euclidean geometry where, in flat space, we can imagine transporting a vector around in space by moving the head and the tail of a vector the same distance in the same direction, so that the vector remains parallel to itself at all times. This is the natural definition of parallel in flat space, and indeed does tie the notion of parallel to that of the metric. In a curved space it is not so simple, and we need to think a little harder as to how we might define parallel vectors in a curved space, or space-time.

where $\Gamma^{\nu}{}_{\mu}$ are numbers that depend on the point P. The covariant derivative of $\vec{b}_{\mu}$, written $\nabla\vec{b}_{\mu}$, is then defined by letting $P' \rightarrow P$ (i.e. $\epsilon \rightarrow 0$) and calculating

$$\nabla\vec{b}_{\mu} := \lim_{\epsilon \rightarrow 0} \frac{\left(\vec{c}'_{\mu} - \vec{b}_{\mu}\right)}{\epsilon} = \lim_{\epsilon \rightarrow 0} \frac{\left(c^{\nu}{}_{\mu}(\epsilon) - \delta^{\nu}_{\mu}\right)\vec{b}_{\nu}}{\epsilon} = \Gamma^{\nu}{}_{\mu}\vec{b}_{\nu}.$$

A complication is that this derivative will in general depend on the direction of P' relative to P; we really need a different derivative for each direction. Again, using the directions associated with the different $\vec{b}_{\rho}$ is sufficient to obtain the general case, so we define different $\Gamma^{\nu}{}_{\mu}$ for each ρ,

$$\nabla_{\rho}\vec{b}_{\mu} = \left(\Gamma_{\rho}\right)^{\nu}{}_{\mu}\,\vec{b}_{\nu}, \tag{4.17}$$

where $\left(\Gamma_{\rho}\right)^{\nu}{}_{\mu}$ are the $\Gamma^{\nu}{}_{\mu}$ for the direction ρ. This is the covariant derivative of $\vec{b}_{\mu}$ in the direction of $\vec{b}_{\rho}$. It is a standard convention to write

$$\Gamma^{\nu}_{\rho\mu} = \left(\Gamma_{\rho}\right)^{\nu}{}_{\mu},$$

and the $\Gamma^{\nu}_{\rho\mu}$ are called *connection coefficients*. Note that $\Gamma^{\nu}_{\rho\mu}$ is not a tensor (see problem 4.1). For this reason, we insist on calling the $\Gamma^{\nu}_{\rho\mu}$ connection *coefficients* – they are not the components of a tensor.

We can now define the covariant derivative of any vector in the direction $\vec{b}_{\rho}$. In a coordinate basis,

$$\nabla_{\rho}\vec{W} = \nabla_{\rho}(W^{\mu}\vec{b}_{\mu}) = (\partial_{\rho}W^{\mu})\vec{b}_{\mu} + W^{\mu}(\nabla_{\rho}\vec{b}_{\mu}) = (\partial_{\rho}W^{\nu} + W^{\mu}\Gamma^{\nu}_{\rho\mu})\vec{b}_{\nu}. \tag{4.18}$$

(We use $\partial_{\rho}W^{\mu}$ for the derivative of the components, as these are not vectors.) The components of the covariant derivative of the vector $\vec{W}$ in the direction ρ are defined by

$$\nabla_{\rho}W^{\nu} := \partial_{\rho}W^{\nu} + \Gamma^{\nu}_{\rho\mu}W^{\mu}. \tag{4.19}$$

We have achieved our goal of defining a derivative of $\vec{W}$ without introducing a second vector field.

A common notation, when dealing with components of a vector, is to use a semicolon to denote covariant differentiation and a comma to denote ordinary partial differentiation, so

$$\left(\nabla_{\rho}\vec{W}\right)^{\nu} = W^{\nu}{}_{;\rho} = W^{\nu}{}_{,\rho} + \Gamma^{\nu}_{\rho\mu}W^{\mu}. \tag{4.20}$$

$W^{\nu}{}_{;\rho}$ are then the components of the covariant derivative of $\vec{W}$ in the ρ-direction. This semicolon notation is only used when dealing with components in a coordinate basis, while ∇_{μ} or ∇_{a} is a more general notation that can be used on complete vectors.

We can obtain the covariant derivative of $\vec{W}$ in any specific direction determined by a vector $\vec{V}$ if we so wish, by taking a linear combination of (4.18),

$$\nabla_{\vec{V}}\vec{W} = V^\rho \nabla_\rho \vec{W} \tag{4.21}$$

or, in components,

$$(\nabla_{\vec{V}}\vec{W})^\mu = V^\rho W^\mu{}_{;\rho}. \tag{4.22}$$

Equation (4.19) is the covariant derivative of a vector. To obtain the covariant derivative of a covector, observe that the length squared of a vector is just a function, as is any inner product, so we don't need a covariant derivative for a length squared or an inner product; just regular partial derivatives will do:

$$\partial_\rho(\vec{V}.\vec{W}) = \partial_\rho(V_\mu W^\mu) = (\partial_\rho V_\mu)W^\mu + V_\mu(\partial_\rho W^\mu).$$

We shall define the covariant derivative in the ρ direction of a covector with components V_μ, written $V_{\mu;\rho}$, so that

$$\partial_\rho(V_\mu W^\mu) = (V_{\mu;\rho})W^\mu + V_\mu(W^\mu{}_{;\rho}).$$

This can be achieved by setting

$$V_{\mu;\rho} = \partial_\rho V_\mu - \Gamma^\nu_{\rho\mu}V_\nu \tag{4.23}$$

(note the minus sign), and we shall adopt this as the definition of the covariant derivative of any covector.

This can be generalised to tensors of higher rank by introducing a separate Γ-symbol for each index, with a plus sign for upper indices and a minus sign for lower indices. For example, the covariant derivative of the metric tensor $g_{\mu\nu}(x)$ is

$$g_{\mu\nu;\rho} = \partial_\rho g_{\mu\nu} - \Gamma^\lambda_{\rho\mu}g_{\lambda\nu} - \Gamma^\lambda_{\rho\nu}g_{\mu\lambda}. \tag{4.24}$$

The covariant derivative of a function $f(x)$ is just the gradient

$$\nabla_\mu f = \partial_\mu f.$$

Covariant differentiation behaves the way you would expect when acting on the sum of two vector fields $\vec{V}(x)$ and $\vec{W}(x)$,

$$\nabla_\rho(\vec{V} + \vec{W}) = \nabla_\rho \vec{V} + \nabla_\rho \vec{W},$$

and if a vector field $\vec{W}(x)$ is multiplied by a function $f(x)$,

$$\nabla_\rho(f\vec{W}) = (\partial_\rho f)\vec{W} + f\nabla_\rho\vec{W}.$$

Defining derivatives of vectors, covectors, and more general tensors in curved spaces is not straightforward; there are, in fact, no less than *three* different kinds of derivatives that are useful in various circumstances: Lie derivatives were introduced on page 90, covariant derivatives have been discussed here, and exterior derivatives were mentioned on page 97. Lie derivatives are useful for discussing symmetries of the geometry (see Equation (4.9)), but their definition relies on a choice of a vector field. Exterior derivatives are only useful for completely anti-symmetric covariant tensors, which happens to be all that is needed for electromagnetism. But covariant derivatives are required in a more general setting, and it is covariant derivatives that we will now focus on.

4.5 Connections and Geodesics

A rather special vector field is one for which

$$\nabla_{\vec{W}}\vec{W} = 0 \tag{4.25}$$

or, more generally,

$$\nabla_{\vec{W}}\vec{W} = \zeta\vec{W} \tag{4.26}$$

for some function $\zeta(\tau)$. Equation (4.26) is particularly important when $\vec{W}$ is the tangent vector to a curve.[14] This is a special property; a completely generic curve will not satisfy this condition. Geometrically, this is saying that when the tangent vector $\vec{W}$ is parallel transported along the curve, it remains tangent to the curve. Such a curve might be called an autoparallel curve.

If the curve is defined parametrically by $x^\mu(\tau)$, then its tangent vector has components

$$W^\mu = \frac{dx^\mu}{d\tau}.$$

and the derivative along the direction of the curve is

---

[14] $\vec{W}$ need not be a complete vector field; Equation (4.26) makes sense when $\vec{W}$ is just a tangent vector to a single curve, although it must be defined at all points along the curve; it is not necessary that it be defined at points off the curve.

$$(\nabla_{\vec{W}}\vec{W})^{\mu} = W^{\nu}W^{\mu}{}_{;\nu} = \frac{dx^{\nu}}{d\tau}\left\{\frac{\partial}{\partial x^{\nu}}\left(\frac{dx^{\mu}}{d\tau}\right) + \Gamma^{\mu}_{\nu\rho}\frac{dx^{\rho}}{d\tau}\right\}$$

$$= \frac{d^2x^{\mu}}{d\tau^2} + \Gamma^{\mu}_{\nu\rho}\frac{dx^{\nu}}{d\tau}\frac{dx^{\rho}}{d\tau}$$

$$= \frac{dW^{\mu}}{d\tau} + \Gamma^{\mu}_{\nu\rho}W^{\nu}W^{\rho}. \tag{4.27}$$

If we choose a different parameterisation of the curve, say $x^{\mu}(\tilde{\tau})$ with $\tilde{\tau}(\tau)$ a function of τ, then the components are

$$\widetilde{W}^{\mu} = \frac{dx^{\mu}}{d\tilde{\tau}} = \frac{d\tau}{d\tilde{\tau}}\frac{dx^{\mu}}{d\tau} \quad\text{and}\quad W^{\mu} \to \widetilde{W}^{\mu} = \frac{d\tau}{d\tilde{\tau}}W^{\mu},$$

so

$$(\nabla_{\vec{W}}\vec{W})^{\mu} \to \frac{d}{d\tilde{\tau}}\widetilde{W}^{\mu} + \Gamma^{\mu}_{\nu\rho}\widetilde{W}^{\nu}\widetilde{W}^{\rho}$$

$$= \frac{d}{d\tilde{\tau}}\left\{\left(\frac{d\tau}{d\tilde{\tau}}\right)W^{\mu}\right\} + \left(\frac{d\tau}{d\tilde{\tau}}\right)^2 \Gamma^{\mu}_{\nu\rho}W^{\nu}W^{\rho}$$

$$= \left(\frac{d^2\tau}{d\tilde{\tau}^2}\right)W^{\mu} + \left(\frac{d\tau}{d\tilde{\tau}}\right)^2\frac{dW^{\mu}}{d\tau} + \left(\frac{d\tau}{d\tilde{\tau}}\right)^2 \Gamma^{\mu}_{\nu\rho}W^{\nu}W^{\rho}$$

$$= \frac{d^2\tau}{d\tilde{\tau}^2}\frac{dx^{\mu}}{d\tau} + \left(\frac{d\tau}{d\tilde{\tau}}\right)^2\left\{\frac{d^2x^{\mu}}{d\tau^2} + \Gamma^{\mu}_{\nu\rho}\frac{dx^{\nu}}{d\tau}\frac{dx^{\rho}}{d\tau}\right\}$$

$$= \left(\frac{d\tau}{d\tilde{\tau}}\right)^2\left\{\frac{d^2x^{\mu}}{d\tau^2} + \Gamma^{\mu}_{\nu\rho}\frac{dx^{\nu}}{d\tau}\frac{dx^{\rho}}{d\tau} + \left(\frac{d\tilde{\tau}}{d\tau}\right)^2\frac{d^2\tau}{d\tilde{\tau}^2}\frac{dx^{\mu}}{d\tau}\right\}.$$

If the tangent vector to the curve satisfies (4.26), then

$$\frac{d^2x^{\mu}}{d\tau^2} + \Gamma^{\mu}_{\nu\rho}\frac{dx^{\nu}}{d\tau}\frac{dx^{\rho}}{d\tau} = \zeta(\tau)\frac{dx^{\mu}}{d\tau},$$

where ζ is a function on the curve. Thus, if we choose $\tilde{\tau}(\tau)$ such that

$$\left(\frac{d\tilde{\tau}}{d\tau}\right)^2\frac{d^2\tau}{d\tilde{\tau}^2} = -\zeta$$

along the curve, we can turn (4.26) into (4.25), $\nabla_{\vec{\tilde{W}}}\vec{\tilde{W}} = 0$, and they are equivalent. We shall therefore focus on (4.25), as it is the simpler of the two.

As has been emphasised, there are two different, in principle independent, concepts in differential geometry: a metric and a connection. These can be, and usually are, related in a very natural way by defining the connection in terms of the metric so that the special curve in Figure 4.3, defined by (4.25), is in fact a geodesic.

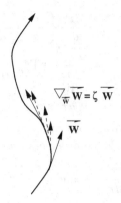

$$\nabla_{\overline{W}} \, \overline{W} = \zeta \, \overline{W}$$

$$\overline{W}$$

Figure 4.3 **Parallel transport along a curve.** A special curve is one for which a vector tangent to a curve remains tangent to the curve when parallel transported in the direction of the curve.

With $W^\mu = \frac{dx^\mu}{d\tau} = \dot{x}^\mu$, Equation (4.25) reads

$$\ddot{x}^\mu + \Gamma^\mu_{\nu\rho} \dot{x}^\nu \dot{x}^\rho = 0. \tag{4.28}$$

Compare this to the geodesic equation following from the action

$$S[x; \dot{x}] = \int_{\tau_0}^{\tau_1} \mathcal{L}(x, \dot{x}) d\tau,$$

with

$$\mathcal{L} = g_{\mu\nu}(x) \dot{x}^\mu \dot{x}^\nu.$$

The equations of motion are

$$\frac{d}{d\tau} \left\{ g_{\mu\nu}(x) \dot{x}^\nu + g_{\nu\mu}(x) \dot{x}^\nu \right\} = \left\{ \partial_\mu g_{\nu\rho}(x) \right\} \dot{x}^\nu \dot{x}^\rho.$$

Since $g_{\mu\nu} = g_{\nu\mu}$ is symmetric and has no explicit τ-dependence, so $\frac{d}{d\tau} = \dot{x}^\rho \partial_\rho$ when acting on $g_{\mu\nu}(x(\tau))$, we can write this as

$$2 g_{\mu\nu}(x) \ddot{x}^\nu + \left(\partial_\rho g_{\mu\nu} + \partial_\rho g_{\nu\mu} - \partial_\mu g_{\nu\rho} \right) \dot{x}^\nu \dot{x}^\rho = 0.$$

Treating $g_{\mu\nu}$ as the components of a square matrix, we can multiply this equation with the inverse matrix $\left(g^{-1} \right)^{\lambda\mu}$ (assuming $\det g \neq 0$) and re-arrange the indices to obtain

$$\ddot{x}^\mu + \frac{1}{2} \left(g^{-1} \right)^{\mu\lambda} \left(\partial_\nu g_{\lambda\rho} + \partial_\rho g_{\nu\lambda} - \partial_\lambda g_{\nu\rho} \right) \dot{x}^\nu \dot{x}^\rho = 0.$$

Now observe the similarity of this equation with (4.28). They are the same if we identify

$$\Gamma^\mu_{\nu\rho} = \frac{1}{2} \left(g^{-1} \right)^{\mu\lambda} \left(\partial_\nu g_{\lambda\rho} + \partial_\rho g_{\nu\lambda} - \partial_\lambda g_{\nu\rho} \right). \tag{4.29}$$

Requiring that (4.25) is the geodesic equation relates the connection coefficients to first partial derivatives of the metric.

This is a natural choice for the connection coefficients, but it does not yet determine them uniquely; it only determines the symmetric part of $\Gamma^\mu_{\nu\rho}$ in the lower two indices, $\frac{1}{2}\left(\Gamma^\mu_{\nu\rho} + \Gamma^\mu_{\rho\nu}\right)$. To get a unique connection we need to specify the anti-symmetric part of $\Gamma^\mu_{\nu\rho}$, which we shall denote by $\tau^\mu_{\nu\rho}$,

$$\tau^\mu_{\nu\rho} := \Gamma^\mu_{\nu\rho} - \Gamma^\mu_{\rho\nu} = 2\Gamma^\mu_{[\nu\rho]}. \tag{4.30}$$

This is a tensor (see problem 4.1), and it is called the *torsion* associated with the connection. There are flavours of general relativity in which torsion plays a role, but in the simplest versions it is set to zero. There is no observational evidence for non-zero torsion in the description of gravity, so from here on we shall assume that $\Gamma^\mu_{\nu\rho} = \Gamma^\mu_{\rho\nu}$ and (4.29) is then called the *Levi–Civita connection*.

As a matter of notation, we mention that the components of the inverse of the metric $\left(g^{-1}\right)^{\mu\nu}$ tend to appear so often in our formulae that it is convenient to adopt a shorthand notation and write

$$g^{\mu\nu} = \left(g^{-1}\right)^{\mu\nu}.$$

At first sight this may seem confusing: how can we tell the metric apart from its inverse if we use the same symbol for both? It was Einstein who realised that this notation can be used without ambiguity. The position of the indices tells us whether we are dealing with the metric tensor itself, $g_{\mu\nu}$, or its inverse, $g^{\mu\nu}$. This notation implies that $g^{\mu\rho}g_{\rho\nu} = \delta^\mu_\nu$ for *any* metric. We shall adopt this convention from now on and write the symmetric part of the connection coefficients as

$$\boxed{\Gamma^\mu_{\nu\rho} = \frac{1}{2}g^{\mu\lambda}\left(\partial_\nu g_{\lambda\rho} + \partial_\rho g_{\nu\lambda} - \partial_\lambda g_{\nu\rho}\right).} \tag{4.31}$$

This is a very important equation, so we have put a box around it. These $\Gamma^\mu_{\nu\rho}$ are also called the *Christoffel symbols* for the Levi–Civita connection associated with the metric $g_{\mu\nu}$. Equation (4.31) relates the connection to the metric; we are defining the connection by demanding that tangent vectors to geodesics satisfy (4.25); with this choice, Equation (4.25) is called the *geodesic equation*.[15]

---

[15] We stress that this is not essential; it is possible to define consistent geometries with different connections, but the Levi–Civita connection is the simplest possibility when we have a well-defined metric, and it is the one that is used in the theory of general relativity.

A nice feature of the connections defined in (4.29) is that the covariant derivative of the metric tensor vanishes:

$$g_{\mu\nu;\rho} = \partial_\rho g_{\mu\nu} - \Gamma^\sigma_{\mu\rho} g_{\sigma\nu} - \Gamma^\sigma_{\nu\rho} g_{\mu\sigma} = 0. \tag{4.32}$$

(Note that $g_{\mu\nu;\rho} = 0 \Rightarrow g^{\mu\nu}{}_{;\rho} = 0$.) In fact, assuming (4.32) is another way of deriving (4.31). Equation (4.32) has the interpretation that the definition of lengths of vectors does not change under parallel transport – parallel transport using the connection coefficients (4.31) preserves lengths; these connection coefficients are metric compatible by construction. For any given metric, the Levi–Civita connection is the unique metric-compatible torsion-free connection.

As an example, consider again the 2-dimensional sphere (2.9) with

$$g_{\mu\nu} = a^2 \begin{pmatrix} 1 & 0 \\ 0 & \sin^2\theta \end{pmatrix}, \qquad g^{\mu\nu} = \frac{1}{a^2} \begin{pmatrix} 1 & 0 \\ 0 & \frac{1}{\sin^2\theta} \end{pmatrix}.$$

There are $2^3 = 8$ connection coefficients and, using (4.31), these are[16]

$$\Gamma^\theta_{\theta\theta} = \Gamma^\theta_{\theta\phi} = \Gamma^\theta_{\phi\theta} = \Gamma^\phi_{\theta\theta} = \Gamma^\phi_{\phi\phi} = 0$$

$$\Gamma^\theta_{\phi\phi} = \frac{1}{2}(-\partial_\theta \sin^2\theta) = -\sin\theta\cos\theta \tag{4.33}$$

$$\Gamma^\phi_{\theta\phi} = \Gamma^\phi_{\phi\theta} = \frac{1}{2\sin^2\theta}(\partial_\theta \sin^2\theta) = \cot\theta.$$

For example,

$$\Gamma^\theta_{\theta\theta} = \frac{1}{2}g^{\theta\alpha}(\partial_\theta g_{\alpha\theta} + \partial_\theta g_{\theta\alpha} - \partial_\alpha g_{\theta\theta}) = 0,$$

and, since $g^{\mu\nu}$ is diagonal, $g^{\theta\alpha} = 0$ unless $\alpha = \theta$; but when $\alpha = \theta$, all the derivatives in $(\partial_\theta g_{\alpha\theta} + \partial_\theta g_{\theta\alpha} - \partial_\alpha g_{\theta\theta})$ vanish because $g_{\theta\theta}$ is a constant. For $\Gamma^\theta_{\phi\phi}$, since no metric components depend on ϕ, $\partial_\phi g_{\mu\nu} = 0$ and

$$\Gamma^\theta_{\phi\phi} = \frac{1}{2}g^{\theta\alpha}(\partial_\phi g_{\alpha\phi} + \partial_\phi g_{\phi\alpha} - \partial_\alpha g_{\phi\phi})$$

$$= -\frac{1}{2}g^{\theta\theta}\partial_\theta g_{\phi\phi} = -\frac{1}{2a}(2a\sin\theta\cos\theta) = -\sin\theta\cos\theta.$$

The calculation of the other connection coefficients is left as an exercise.

The coordinate basis vectors, $\vec{b}_\theta = \frac{\partial}{\partial\theta}$, $\vec{b}_\phi = \frac{\partial}{\partial\phi}$, are parallel transported according to

---

[16] Note that the overall length scale, the radius of the sphere a, drops out of the connection. This is always the case; the connection only cares about angles and relative lengths, not the overall scale.

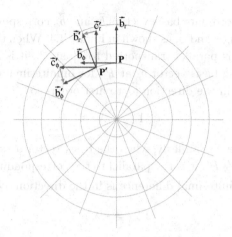

Figure 4.4 **Parallel transport and polar coordinates in two dimensions.** Parallel transporting basis vectors $\vec{b}_\alpha$ from P to $\vec{c}\,'_\alpha$ at P', they do not correspond exactly to the basis vectors $\vec{b}\,'_\alpha$ at P'. The connection coefficients (4.35) encode this difference (short arrows).

$$\nabla_\theta\,\vec{b}_\theta = \Gamma^\theta_{\theta\theta}\,\vec{b}_\theta + \Gamma^\phi_{\theta\theta}\,\vec{b}_\phi = 0,$$

$$\nabla_\theta\,\vec{b}_\phi = \nabla_\phi\vec{b}_\theta = \Gamma^\theta_{\theta\phi}\,\vec{b}_\theta + \Gamma^\phi_{\theta\phi}\,\vec{b}_\phi = \cot\theta\,\vec{b}_\phi, \qquad (4.34)$$

$$\nabla_\phi\vec{b}_\phi = \Gamma^\theta_{\phi\phi}\,\vec{b}_\theta + \Gamma^\phi_{\phi\phi}\,\vec{b}_\phi = -\cos\theta\sin\theta\,\vec{b}_\theta.$$

These allow us to calculate how orthonormal vectors

$$\vec{b}_1 = \frac{1}{a}\frac{\partial}{\partial\theta}, \qquad \vec{b}_2 = \frac{1}{a\sin\theta}\frac{\partial}{\partial\phi}$$

should be parallel transported in the $\vec{b}_1$ and $\vec{b}_2$ directions, for example,

$$\nabla_1\vec{b}_2 = \frac{1}{a}\nabla_\theta\left\{\left(\frac{1}{a\sin\theta}\right)\vec{b}_\phi\right\} = \frac{1}{a^2}\left(\frac{1}{\sin\theta}\nabla_\theta\,\vec{b}_\phi + \frac{\partial}{\partial\theta}\left(\frac{1}{\sin\theta}\right)\vec{b}_\phi\right) = 0.$$

The other combinations work out to be

$$\nabla_1\vec{b}_1 = 0, \qquad \nabla_2\vec{b}_1 = \frac{\cot\theta}{a}\vec{b}_1, \qquad \nabla_2\vec{b}_2 = -\frac{\cot\theta}{a}\vec{b}_1.$$

Note that $\nabla_1\vec{b}_2 \neq \nabla_2\vec{b}_1$.

Connection coefficients are not only important in curved spaces; they are also necessary if we are using curvilinear coordinates in flat space. Consider the line element for the 2-dimensional Euclidean plane in polar coordinates:

$$ds^2 = dr^2 + r^2 d\phi^2.$$

We can use coordinate basis vectors $\vec{b}_r$ and $\vec{b}_\phi$ corresponding to directions of increasing r and ϕ, as shown in Figure 4.4. When the basis vector $\vec{b}_r$ at a point P is parallel transported to $\vec{c}'_r$ at P', it is not parallel to the corresponding basis vector $\vec{b}'_r$ at P'; for infinitesimal displacements, the difference is in the direction $\vec{b}_\phi$ to first order, so

$$\Gamma^{\phi}_{r\phi} = \lambda$$

for some positive λ. Similarly, when the basis vector $\vec{b}_\phi$ at P is parallel transported to $\vec{c}'_\phi$ P', it is not parallel to the corresponding basis vector $\vec{b}'_\phi$ at P'; the infinitesimal difference is in the direction $-\vec{b}_r$, so

$$\Gamma^{r}_{\phi\phi} = -\tilde{\lambda}$$

for some positive $\tilde{\lambda}$. The further P and P' are away from each other, the bigger the discrepancy; but for infinitesimal distances, (4.31) can be used to calculate the connection coefficients and show that $\lambda = \tilde{\lambda} = r$:

$$\Gamma^{\phi}_{r\phi} = \frac{1}{r} \qquad \Gamma^{r}_{\phi\phi} = -r. \tag{4.35}$$

Yet again, we emphasise that the connection coefficients $\Gamma^{\mu}_{\nu\rho}$ are *not* the components of a tensor. They do not transform like a tensor. Also, remember the discussion in §2.3 (page 21) when we use x-y coordinates near the N pole of a sphere: any curved space (or space-time) looks flat in any small-enough region round any given point P, so we can always choose coordinates which look like Cartesian coordinates at P, as long as we don't stray too far from P. In these coordinates the connection coefficients $\Gamma^{\mu}_{\nu\rho}$ vanish at P.[17] If space is curved, the $\Gamma^{\mu}_{\nu\rho}$ will inevitably become non-zero as we move away from P – only if the space is flat is it possible to find Cartesian coordinates in which the $\Gamma^{\mu}_{\nu\rho}$ vanish everywhere.

This has an extremely important consequence for general relativity: it is always possible to choose coordinates at a given event P so that $\Gamma^{\mu}_{\nu\rho}$ vanish at P, though not necessarily at any other point.[18] In these coordinates (locally inertial coordinates) space-time looks 'locally' like Minkowski space-time to a freely falling observer at P (a locally inertial observer). If there is a gravitational field present, such an observer is

---

[17] This can be done because $\Gamma^{\mu}_{\nu\rho}$ is not a tensor. If the components of a tensor are not all zero at some point, they cannot all be made to vanish at that point by a coordinate transformation.

[18] The unique exception is flat space-time in which the connection coefficients vanish everywhere if we are using Cartesian coordinates.

in free fall (like the observer in the lift in Section 1.1). Any deviation from flat Minkowski space-time (tidal forces) is encoded in connection coefficients $\Gamma^\mu_{\nu\rho}$ which become non-zero away from P. The mathematical statement of the relativistic map maker's problem described in §2.8 is that, if space-time is curved, the connection coefficients cannot be made to vanish everywhere; there is no global inertial coordinate system unless space-time is Minkowski.

Thus, there is information about the curvature in the connection coefficients, but they are an unreliable witness because we can always choose coordinates so that they vanish at any point, and they may be non-zero even if space is flat (if we use curvilinear coordinates), so we can be fooled. In the next section we describe a more reliable way of determining whether or not a given space is curved.

4.6 The Riemann Tensor

A unique characteristic of curvature is that if a vector is parallel transported around a closed loop and brought back to the starting point, then it will in general be rotated if space is curved. Consider, for example, transporting a vector around a large triangle on the surface of a sphere as shown in Figure 4.5.

Choose the three sides in the figure to be great circles, so they are geodesics, and use the Levi–Civita connection (4.31) so that a tangent vector to a geodesic remains tangent to the geodesic when parallel transported along it. Also, a vector perpendicular to the equator remains perpendicular to the equator when parallel transported along it. When the parallel transported vector returns to the starting point, it has rotated through the angle α. A rotation like this is the smoking gun of curvature; there is no rotation in flat space, even if curvilinear coordinates are used and the connection coefficients do not vanish. For a space-time it could be a boost or a more general Lorentz transformation (a combination of a boost and a rotation) rather than just a simple rotation; we might call this a 'generalised rotation', but for simplicity we shall just use 'rotation' in the following discussion.

Consider a parallelogram, generated by two vectors $\overrightarrow{U}$ and $\overrightarrow{V}$ as in Figure 4.6, rather than a more general loop. First suppose that $\overrightarrow{U} = \frac{\partial}{\partial x^\rho}$ and $\overrightarrow{V} = \frac{\partial}{\partial x^\sigma}$ are basis vectors associated with a specific coordinate system; this is sufficient to understand the general case. In a curved space, a general vector $\overrightarrow{W}$ parallel transported around a parallelogram defined by $\frac{\partial}{\partial x^\rho}$ and $\frac{\partial}{\partial x^\sigma}$ is rotated relative to its original direction when returned

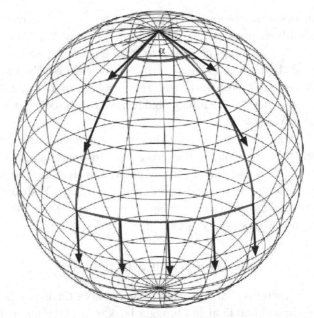

Figure 4.5 **Parallel transport around a spherical triangle.** Starting at the N pole, when a vector is parallel transported around a closed curve and brought back to the N pole, the length has not changed, but the vector has rotated due to the curvature.

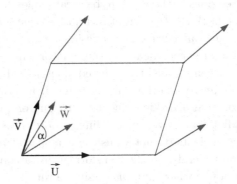

Figure 4.6 **Transporting a vector around a parallelogram.** When a vector $\vec{W}$ is parallel transported around a parallelogram defined by two vectors $\vec{U}$ and $\vec{V}$, $\vec{W}$ is rotated through an angle α determined by the area of the parallelogram and by the Riemann tensor.

to the starting point through an angle α. Suppose the parallelogram is defined by parallel transporting first by $\epsilon\nabla_\rho$, then by $\epsilon\nabla_\sigma$ and back again, where is ϵ is small. (We need to use ∇_ρ and ∇_σ rather than just partial derivatives $\frac{\partial}{\partial x^\rho}$ and $\frac{\partial}{\partial x^\sigma}$ because we are parallel transporting vectors, and that necessarily requires introducing the connection coefficients.) The amount of rotation will depend on the directions of two vectors,[19] and it is shown in Appendix B (Equation (B.2)) that it also depends on ϵ^2:

$$W^\mu \quad \longrightarrow \quad W^\mu + \epsilon^2 \left([\nabla_\sigma, \nabla_\rho]\overrightarrow{W}\right)^\mu + O(\epsilon^3). \tag{4.36}$$

While it is not immediately obvious, the components of $[\nabla_\sigma, \nabla_\rho]\overrightarrow{W}$ are actually an algebraic linear combination of the components of $\overrightarrow{W}$, which we denote by

$$\left([\nabla_\sigma, \nabla_\rho]\overrightarrow{W}\right)^\mu = W^\mu{}_{;\rho;\sigma} - W^\mu{}_{;\sigma;\rho} = R^\mu{}_{\nu\sigma\rho}W^\nu.$$

Explicit evaluation of the covariant derivatives (details are given in Appendix B) shows that

$$R^\mu{}_{\nu\sigma\rho} = \partial_\sigma \Gamma^\mu_{\nu\rho} - \partial_\rho \Gamma^\mu_{\nu\sigma} + \Gamma^\mu_{\lambda\sigma}\Gamma^\lambda_{\nu\rho} - \Gamma^\mu_{\lambda\rho}\Gamma^\lambda_{\nu\sigma}. \tag{4.37}$$

$R^\mu{}_{\nu\sigma\rho}$ is a tensor called the *Riemann tensor*. Naïvely it has D^4 components in D dimensions, since each of the four indices can take one of D possible values. That would be $4^4 = 256$ in four dimensions, but we shall see that it actually has fewer independent components than this.

Equation (4.36) shows us that when we parallel transport a vector W^μ around a parallelogram in the ρ-σ plane, it changes by an amount,

$$\delta W^\mu = \epsilon^2 R^\mu{}_{\nu\sigma\rho}W^\nu,$$

proportional to ϵ^2 and to the Riemann tensor $R^\mu{}_{\nu\sigma\rho}$. Now ϵ^2 is proportional to the area of the parallelogram,[20] and $R^\mu{}_{\nu\sigma\rho}W^\nu$ in the preceding formula represents a rotation (or a Lorentz transformation) of the vector W^μ in the ρ-σ plane. There is no rotation if the space is flat, in which case $R^\mu{}_{\nu\sigma\rho}$ vanishes. Conversely, the space is not flat if $R^\mu{}_{\nu\sigma\rho}$ does not vanish. For a fixed ϵ, the larger the rotation, the greater the curvature (or tidal forces in space-time). This is what we have been looking for – it is the Riemann tensor that describes the degree of curvature of the space and tidal forces in space-time.

---

[19] The sign depends on which way we go around the parallelogram.

[20] Which is $\epsilon^2 \sqrt{|g_{\rho\rho}g_{\sigma\sigma} - g_{\sigma\rho}^2|}$.

In summary:

- Two vector fields can be used to define infinitesimal parallelograms at every point in space, if they are not parallel at the point.
- If the metric is not flat, then parallel transport of a third vector around the parallelogram will result in a rotation of that vector through an angle proportional to the area of the parallelogram and the degree of curvature of the space.

In equations (with a re-labelling of the indices):

$$W^\rho{}_{;\nu;\mu} - W^\rho{}_{;\mu;\nu} = R^\rho{}_{\lambda\mu\nu}W^\lambda \quad (4.38a)$$
$$R^\rho{}_{\lambda\mu\nu} = \partial_\mu\Gamma^\rho_{\lambda\nu} - \partial_\nu\Gamma^\rho_{\lambda\mu} + \Gamma^\rho_{\sigma\mu}\Gamma^\sigma_{\lambda\nu} - \Gamma^\rho_{\sigma\nu}\Gamma^\sigma_{\lambda\mu}. \quad (4.38b)$$

Less complete information about the curvature is contained in the *Ricci tensor*,

$$R_{\mu\nu} = R^\rho{}_{\mu\rho\nu}, \quad (4.39)$$

and the trace of the Ricci tensor, the *Ricci scalar*,

$$R = g^{\mu\nu}R_{\mu\nu}. \quad (4.40)$$

To reiterate: the net effect of parallel transporting $\vec{W}$ around a parallelogram defined by two coordinate basis vectors $\vec{b}_\mu$ and $\vec{b}_\nu$ is to come back to where we started, but to rotate $\vec{W}$ by an amount proportional to the area of the parallelogram. After dividing by the area of the parallelogram, the ϵ^2 disappears and the residual rotation is a measure of the true curvature of the space. This rotation is called *holonomy*, and Figure 4.6 is an attempt to represent this.

There is no rotation and no such holonomy in flat space – the Riemann tensor completely characterises the curvature of space.

Space (space-time) is flat if and only if the Riemann tensor vanishes everywhere.

Some important properties of the Riemann tensor derived from the Levi–Civita connection[21] are derived in Appendix B and we list them here:

---

[21] These are not all completely general in differential geometry; some are only true when the torsion vanishes.

- It is anti-symmetric under interchange of the last two indices:

$$R^{\mu}{}_{\nu\rho\lambda} = -R^{\mu}{}_{\nu\lambda\rho}. \tag{4.41}$$

- It vanishes under complete anti-symmetrisation of the last three indices:

$$R^{\mu}{}_{[\nu\rho\lambda]} = 0. \tag{4.42}$$

- If the first index is lowered using the metric, to change it from a tensor of type[22] $(1,3)$ to a tensor of type $(0,4)$ so that $R_{\mu\nu\rho\lambda} := g_{\mu\sigma}R^{\sigma}{}_{\nu\rho\lambda}$, it is symmetric under interchange of the first and last pair of indices:

$$R_{\mu\nu\rho\lambda} = R_{\rho\lambda\mu\nu}. \tag{4.43}$$

- A consequence of (4.41) and (4.43) is that the Ricci tensor is symmetric:

$$R_{\mu\nu} = R_{\nu\mu}. \tag{4.44}$$

- Another consequence of (4.41) and (4.43) is that the Riemann tensor, when written as a tensor of type (0,4), is also anti-symmetric in the first two indices:[23]

$$R_{\mu\nu\rho\lambda} = -R_{\nu\mu\rho\lambda}. \tag{4.45}$$

- The covariant derivative of the Riemann tensor anti-symmetrised on three indices vanishes:

$$R_{\mu\nu[\rho\lambda;\sigma]} = 0. \tag{4.46}$$

Number of Independent Components of $R^{\mu}{}_{\nu\rho\lambda}$
The symmetries (4.41)–(4.46) reduce the number of independent components of the Riemann tensor. In four dimensions, (4.41) reduces the number of independent possibilities for the pair of indices $\{\rho\lambda\}$ from $4 \times 4 = 16$ to 6 (the number of components in an anti-symmetric 4×4 matrix), while (4.43), together with (4.41), implies that

$$R_{\mu\nu\rho\lambda} = R_{\rho\lambda\mu\nu} = -R_{\rho\lambda\nu\mu} = -R_{\nu\mu\rho\lambda},$$

which also reduces the number of independent possibilities for $\{\mu\nu\}$ from 16 to 6. Using (4.43) again, we can now view $R_{[\mu\nu][\rho\lambda]}$ as a symmetric

---

[22] A tensor of type (p,q) has p indices as superscripts and q indices as subscripts; see Appendix A.
[23] This is essentially the reason that we can interpret $R^{\mu}{}_{\nu\rho\sigma}W^{\nu}$ as being related to a rotation in the ρ-σ plane – anti-symmetric matrices generate rotations.

6×6 matrix, which has only $\frac{1}{2}.7.6 = 21$ components, a huge reduction from 256. We can always write

$$R_{\mu\nu\rho\lambda} = R_{[\mu\nu][\rho\lambda]},$$

but the square brackets are redundant if we remember that the Riemann tensor is anti-symmetric in both its first and last pair of indices.

Lastly, Equation (4.42) gives more constraints. In four dimensions there are four independent combinations of the anti-symmetric triple $[\nu\rho\lambda]$. Lowering the first index on $R^{\mu}{}_{\nu\rho\lambda}$ and using (4.45), we can write (4.42) as

$$R_{\mu\nu\rho\lambda}\varepsilon^{\nu\rho\lambda\sigma} = R_{[\mu\nu]\rho\lambda}\varepsilon^{\nu\rho\lambda\sigma} = 0,$$

where $\varepsilon^{\nu\rho\lambda\sigma}$ is the 4-dimensional ε-tensor,[24] but now we are anti-symmetrising over the first two and the last three indices on $R_{\mu\nu\rho\lambda}$, which is equivalent to anti-symmetrising over all four indices,

$$R_{[\mu\nu\rho\lambda]} = 0.$$

In four dimensions there is only one possible choice of four anti-symmetrised indices; there is only one independent constraint in (4.42) in four dimensions and

$$R_{\mu[\nu\rho\lambda]} = 0 \qquad \Rightarrow \qquad \epsilon^{\mu\nu\rho\lambda}R_{\mu\nu\rho\lambda} = 0.$$

In summary, the Riemann tensor has a total of 20 independent components in four dimensions.

Equation (4.46) gives a differential relation, not an algebraic one. It gives constraints between the covariant derivatives of the Riemann tensor and can be used to show that the Ricci tensor satisfies

$$R^{\mu}{}_{\nu;\mu} = \frac{1}{2}\partial_{\nu}R$$

(see Appendix B.2). If we define

$$G_{\mu\nu} = R_{\mu\nu} - \frac{1}{2}g_{\mu\nu}R, \tag{4.47}$$

then we have

$$G^{\mu\nu}{}_{;\mu} = 0, \tag{4.48}$$

since $g_{\nu\rho;\mu} = g^{\nu\rho}{}_{;\mu} = 0$. Equations (4.47) and (4.48) will be crucial for deriving Einstein's equations in the next section.

---

[24] $\varepsilon^{0123} = +1$ and $\varepsilon^{\mu\nu\rho\sigma}$ is completely anti-symmetric on all four indices, vanishing if any two indices are the same and equal to ± 1 if they all differ.

In two dimensions the Riemann tensor has only one independent component, namely

$$R_{1212},$$

all others vanish. For example, for the 2-dimensional sphere of radius a, with connection coefficients (4.33),

$$\begin{aligned}
R^{\theta}{}_{\phi\theta\phi} &= \partial_{\theta}\Gamma^{\theta}_{\phi\phi} - \partial_{\phi}\Gamma^{\theta}_{\phi\theta} + \Gamma^{\theta}_{\sigma\theta}\Gamma^{\sigma}_{\phi\phi} - \Gamma^{\theta}_{\sigma\phi}\Gamma^{\sigma}_{\phi\theta} \\
&= (\sin^2\theta - \cos^2\theta) - (-\sin\theta\cos\theta)\cot\theta \\
&= \sin^2\theta,
\end{aligned} \tag{4.49}$$

giving

$$R_{\theta\phi\theta\phi} = g_{\theta\lambda}R^{\lambda}{}_{\phi\theta\phi} = a^2\sin^2\theta.$$

The Ricci tensor is diagonal,

$$R_{\theta\theta} = g^{\phi\phi}R_{\theta\phi\theta\phi} = 1, \qquad R_{\phi\phi} = g^{\theta\theta}R_{\phi\theta\phi\theta} = g^{\theta\theta}R_{\theta\phi\theta\phi} = \sin^2\theta;$$

in fact,

$$R_{\alpha\beta} = \frac{1}{a^2}g_{\alpha\beta} = \begin{pmatrix} 1 & 0 \\ 0 & \sin^2\theta \end{pmatrix}. \tag{4.50}$$

The Ricci scalar is

$$R = g^{\alpha\beta}R_{\alpha\beta} = \frac{2}{a^2}.$$

Since the Riemann tensor has only one independent component in two dimensions, the Ricci scalar actually has complete information about the curvature – for a sphere it is constant and inversely proportional to the square of the radius of the sphere. This is natural; spherical symmetry dictates that the curvature is the same everywhere and the same in all directions, and the curvature gets smaller as the radius of the sphere increases. It is always the case that the Ricci scalar has dimensions of (*inverse length*)^2.

Examples of the Riemann tensor for some other metrics are given in the appendices. For the moment we repeat, with emphasis:

> **The metric on any space(-time) is flat if and only if the Riemann tensor vanishes everywhere.**

This is the only surefire criterion for determining whether or not a given metric is flat. In 4-dimensional space-time, there are tidal forces if and only if $R^{\mu}{}_{\nu\rho\sigma} \neq 0$.

In visualising curved spaces, it is inevitable that one pictures a curved surface embedded in flat 3-dimensional space. But there is a really beautiful mathematical result due to Gauss, his *Theorema Egregium* (Remarkable Theorem), that curvature is actually intrinsic to the space itself; it does not depend on any embedding into some higher-dimensional space. Gauss originally proved this for 2-dimensional surfaces, but it is true in any dimension. The Riemann tensor is purely intrinsic to the space under study; it does not rely on any embedding into a higher-dimensional space.

Summary

This chapter has inevitably been somewhat technical, as the mathematics for describing curved spaces is rather abstract. Below, the important equations that are needed to understand the remaining chapters are summarised:

$$\Gamma^\mu_{\nu\rho} = \Gamma^\mu_{\rho\nu} = \frac{1}{2}g^{\mu\lambda}\left(\partial_\nu g_{\lambda\rho} + \partial_\rho g_{\nu\lambda} - \partial_\lambda g_{\nu\rho}\right) \qquad \text{Levi–Civita connection}$$

$$R^\mu{}_{\nu\rho\sigma} = \partial_\rho \Gamma^\mu_{\nu\sigma} - \partial_\sigma \Gamma^\mu_{\nu\rho} + \Gamma^\mu_{\lambda\rho}\Gamma^\lambda_{\nu\sigma} - \Gamma^\mu_{\lambda\sigma}\Gamma^\lambda_{\nu\rho} \qquad \text{Riemann tensor}$$

$$R_{\mu\nu} = R_{\nu\mu} = R^\rho{}_{\mu\rho\nu} \qquad \text{Ricci tensor}$$

$$R = R^\mu{}_\mu = g^{\mu\nu}R_{\mu\nu}, \qquad \text{Ricci scalar}$$

$$G_{\mu\nu} = G_{\nu\mu} = R_{\mu\nu} - \frac{1}{2}g_{\mu\nu}R \qquad \text{Einstein tensor}$$

$$G^{\mu\nu}{}_{;\nu} = \partial_\nu G^{\mu\nu} + \Gamma^\mu_{\rho\nu}G^{\rho\nu} + \Gamma^\nu_{\rho\nu}G^{\mu\rho} = 0 \qquad \text{Bianchi identity}$$

4.7 Problems

1) Under a general coordinate transformation

$$dx^{\mu'} = \frac{\partial x^{\mu'}}{\partial x^\nu}dx^\nu$$

and

$$\vec{b}_\mu \to \vec{b}_{\mu'} = \left(\frac{\partial x^\nu}{\partial x^{\mu'}}\right)\vec{b}_\nu, \tag{4.51}$$

show that the Christoffel symbols (4.31) transform as

$$\Gamma^{\mu'}_{\nu'\tau'} = \frac{\partial x^{\mu'}}{\partial x^\rho}\frac{\partial x^\lambda}{\partial x^{\nu'}}\frac{\partial x^\eta}{\partial x^{\tau'}}\Gamma^\rho_{\lambda\eta} + \frac{\partial^2 x^\eta}{\partial x^{\nu'}\partial x^{\tau'}}\frac{\partial x^{\mu'}}{\partial x^\eta}.$$

This shows that $\Gamma^{\rho}_{\lambda\eta}$ is *not* a tensor. However, the ugly term on the right-hand side cancels if we take only the anti-symmetric part

$$\tau^{\nu}_{\mu\rho} = \Gamma^{\nu}_{\mu\rho} - \Gamma^{\nu}_{\rho\mu};$$

the torsion *is* a tensor: it is consistent to set it to zero.

2) Using the expression in Appendix E for the Riemann tensor arising from a general Robertson–Walker metric, show that the Robertson–Walker universe with $K = -1$ and $a = ct$ is just flat Minkowski space-time in an unusual coordinate system.

3) Connections can be obtained by embedding curved spaces in larger-dimensional flat spaces and projecting. Consider a 2-dimensional sphere in 3-dimensional Euclidean space with radius r and points labelled by

$$\vec{r} = (r\cos\phi\sin\theta, r\sin\phi\sin\theta, r\cos\theta)$$

in spherical polar coordinates. The unit normal to the sphere at any point on its surface is

$$\vec{n} = (\cos\phi\sin\theta, \sin\phi\sin\theta, \cos\theta).$$

Show that the vectors

$$\vec{u}_{\theta} = \partial_{\theta}\vec{n} \qquad \text{and} \qquad \vec{u}_{\phi} = \partial_{\phi}\vec{n}$$

are tangent to a sphere of constant radius.

Calculate the derivatives $\partial_{\theta}\vec{u}_{\theta}$, $\partial_{\phi}\vec{u}_{\theta}$, $\partial_{\theta}\vec{u}_{\phi}$, and $\partial_{\phi}\vec{u}_{\phi}$ and show that they are not all tangent to the sphere. But if we now project these vectors onto the tangent plane to the sphere, by eliminating their components normal to the sphere, we get

$$\partial_{\theta}\vec{u}_{\theta} \quad \rightarrow \quad 0,$$
$$\partial_{\phi}\vec{u}_{\phi} \quad \rightarrow \quad -\sin\theta\cos\theta\,\vec{u}_{\theta},$$
$$\partial_{\theta}\vec{u}_{\phi} \quad \rightarrow \quad \cot\theta\vec{u}_{\phi},$$
$$\partial_{\phi}\vec{u}_{\theta} \quad \rightarrow \quad \cot\theta\vec{u}_{\phi}.$$

Referring to (4.34), these are

$$\partial_{\mu}\vec{u}_{\nu} = \Gamma^{\rho}_{\mu\nu}\vec{u}_{\rho}.$$

While this is a very intuitive way of understanding connections geometrically, it is not necessarily physical. In general relativity, curved 4-dimensional space-times are not viewed as being physically embedded in a larger-dimensional flat space-time; their curvature is intrinsic to their own geometry.

4) Work through the details in Appendix B to prove Equations (B.33a)–(B.33c).

5) Show that, if the metric is varied by a small amount

$$g_{\mu\nu} \to g_{\mu\nu} + \delta g_{\mu\nu},$$

a) the change in the connection coefficients is

$$\delta\Gamma^{\mu}_{\nu\rho} = \frac{1}{2}g^{\mu\lambda}\big\{(\delta g_{\nu\lambda})_{;\rho} + (\delta g_{\lambda\rho})_{;\nu} - (\delta g_{\nu\rho})_{;\lambda}\big\},$$

where terms of order δ^2 are ignored.[25] ($\delta\Gamma^{\mu}_{\nu\rho}$ *is a tensor, the difference of two connections is a tensor of type* $(1,2)$.)

b) the change in the Riemann tensor is

$$\delta R^{\mu}{}_{\nu\rho\sigma} = \big(\delta\Gamma^{\mu}_{\nu\sigma}\big)_{;\rho} - \big(\delta\Gamma^{\mu}_{\nu\rho}\big)_{;\sigma}$$

c) for the Ricci tensor,

$$\delta R_{\mu\nu} = \big(\delta\Gamma^{\lambda}_{\mu\nu}\big)_{;\lambda} - \big(\delta\Gamma^{\lambda}_{\mu\lambda}\big)_{;\nu}$$

$$= \frac{1}{2}g^{\lambda\rho}\big\{(\delta g_{\rho\nu})_{;\mu;\lambda} + (\delta g_{\rho\mu})_{;\nu;\lambda} - (\delta g_{\mu\nu})_{;\rho;\lambda} - (\delta g_{\lambda\rho})_{;\mu;\nu}\big\}$$

d) and $\delta R = g^{\mu\lambda}g^{\nu\rho}(\delta g_{\lambda\rho})_{;\nu;\mu} - g^{\lambda\rho}g^{\mu\nu}(\delta g_{\mu\nu})_{;\lambda;\rho} - R^{\mu\nu}\delta g_{\mu\nu}.$

[25] Hint: in this approximation, for

$$g_{\mu\nu} \to g_{\mu\nu} + \delta g_{\mu\nu}$$

the inverse metric components change to

$$g^{\mu\nu} \to g^{\mu\nu} - g^{\mu\lambda}\delta g_{\lambda\tau}g^{\tau\nu},$$

that is, as a matrix $\delta g^{-1} = -g^{-1}(\delta g)g^{-1}$.

5

Einstein's Field Equations

5.1 Derivation of the Equations

If gravitational forces are described by curvature, then the source of that curvature must be the same as the source of the gravitational force, mass or equivalently energy. Einstein suggested that the curvature of space-time is caused by the distribution of mass/energy: energy generates curvature. We must be careful to include all kinds of energy for a complete picture. Because of Einstein's famous formula $E = mc^2$, we should expect matter, with density ρ, to give an energy density ρc^2. But we should also include kinetic energy, or momentum (or momentum density, momentum per unit volume, p^α), in addition to energy arising from internal stresses in matter, be it fluid or solid, described by the pressure tensor $P^{\alpha\beta}$, as well as energy and stresses in electromagnetic fields. These energies are encoded into 10 functions that are naturally arranged into a symmetric 4×4 matrix (see Appendix C),

$$T^{\mu\nu}(x) = \begin{pmatrix} \rho c^2 & p^1 c & p^2 c & p^3 c \\ p^1 c & P^{11} & P^{12} & P^{13} \\ p^2 c & P^{21} & P^{22} & P^{23} \\ p^3 c & P^{31} & P^{32} & P^{33} \end{pmatrix}, \tag{5.1}$$

where $P^{\alpha\beta} = P^{\beta\alpha}$ for $\alpha, \beta = 1, 2, 3$ is the pressure tensor, and $x^0 = ct$.

For a relativistic fluid, for example, with mass density ρ and pressure P, it is shown in Appendix C, Equation (C.14), that

$$T^{\mu\nu} = \left(\rho + \frac{P}{c^2} \right) U^\mu U^\nu + g^{\mu\nu} P, \tag{5.2}$$

where $U^\mu = \gamma(v)(c, v^1, v^2, v^3)$ is the 4-velocity of the fluid. It is also shown in Appendix C that $T_{\mu\nu}$ is covariantly constant in the sense that

$$T^{\mu\nu}{}_{;\nu} = 0.$$

The geometry is described by the metric $g_{\mu\nu}$, which is a 4×4 symmetric matrix satisfying

$$g^{\mu\nu}{}_{;\nu} = 0,$$

and it is tempting to suggest that these might be proportional to each other:

$$g^{\mu\nu} = \kappa T^{\mu\nu}, \tag{5.3}$$

where κ is a constant. However, this does not work: $T^{\mu\nu} = 0$ corresponds to no matter at all, empty space-time, and then space-time should be flat – when there is no matter or energy present, $g^{\mu\nu}$ should be the flat Minkowski metric. This is clearly not possible for any finite κ.

Fortunately, there is another rank-2 symmetric tensor that contains information about the geometry, is covariantly constant, and furthermore it vanishes for the Minkowski metric: the Einstein tensor $G^{\mu\nu}$ in Equation (4.47) (see also (B.31) in Appendix B.2), so we could try relating the geometry to the matter distribution via

$$G^{\mu\nu} = \kappa T^{\mu\nu},$$

or equivalently

$$G_{\mu\nu} = \kappa T_{\mu\nu}. \tag{5.4}$$

Now, if we use coordinates with dimensions of length, $g_{\mu\nu}$ are dimensionless and the entries in $T_{\mu\nu}$ have dimensions of energy density (kg $m^{-1}s^{-2}$), while the entries in $G_{\mu\nu}$ have dimensions of inverse length squared (they are obtained by differentiating $g_{\mu\nu}$ twice), so κ must have dimensions of $kg^{-1}m^{-1}s^2$. We assume that κ involves Newton's constant, which has dimensions $kg^{-1}m^3s^{-2}$, and since we are looking for a relativistic theory of gravity we shall balance the dimensions by using powers of the speed of light, c. We are looking for a combination of powers of G and c with dimensions of $kg^{-1}m^{-1}s^2$; the only possibility is $\frac{G}{c^4}$. We shall therefore assume that

$$G_{\mu\nu} = \frac{kG}{c^4} T_{\mu\nu}, \tag{5.5}$$

where k is a dimensionless number.

The coefficient k can be determined by requiring that the non-relativistic limit of (5.5) reproduce the field equation of Newtonian gravity (1.10),

$$\nabla^2 \Phi = 4\pi G \rho, \tag{5.6}$$

when $c^2 \to \infty$. In this limit, the energy density ρc^2 completely dominates over both the pressure and the momentum density in (5.1) and

$$\frac{1}{c^2} T_{\mu\nu} \longrightarrow \begin{pmatrix} \rho & 0 & 0 & 0 \\ 0 & 0 & 0 & 0 \\ 0 & 0 & 0 & 0 \\ 0 & 0 & 0 & 0 \end{pmatrix} + O\left(\frac{1}{c}\right). \tag{5.7}$$

We saw in §3.5 that the geodesic equation associated with the line element (3.17), for which

$$g_{00} = -\left(1 + \frac{2\Phi}{c^2}\right), \qquad g_{\alpha\beta} = \delta_{\alpha\beta}, \tag{5.8}$$

reproduces (1.7), with $x^0 = ct$. However, simply sending $c \to \infty$ in $G_{00} = \frac{8\pi G}{c^4} T_{00}$ using this line element with (5.7) just gives $0 = 0$, so we need to be more careful about how we take the limit.

Re-express (5.5) in terms of the Ricci tensor $R_{\mu\nu}$ and the Ricci scalar $R = R^\mu{}_\mu$, and then take the trace

$$G_{\mu\nu} = R_{\mu\nu} - \frac{1}{2} g_{\mu\nu} R = \frac{kG}{c^4} T_{\mu\nu} \tag{5.9}$$

$$\Rightarrow \qquad R = -\frac{kG}{c^4} T, \tag{5.10}$$

where $T = T^\mu{}_\mu$ is the trace of $T_{\mu\nu}$. Equation (5.9) can therefore be re-written as

$$c^2 R_{\mu\nu} = \frac{kG}{c^2}\left(T_{\mu\nu} - \frac{1}{2} g_{\mu\nu} T\right). \tag{5.11}$$

All we need to know is the 00 component of the Ricci tensor associated with the metric (5.8). The calculation is left as an exercise; the result is

$$R_{00} = \frac{\nabla^2 \Phi}{c^2} + O\left(\frac{\Phi}{c^2}\right)^2. \tag{5.12}$$

Now using the metric (5.8), the trace of $T_{\mu\nu}$ in (5.7) satisfies

$$T = -\frac{\rho c^2}{\left(1 + \frac{2\Phi}{c^2}\right)} + \cdots = -\rho c^2 + \cdots . \tag{5.13}$$

Putting (5.12) and (5.13) into (5.11), the 00 component satisfies

$$c^2 R_{00} = \nabla^2 \Phi + O\left(\frac{1}{c^2}\right) = \frac{kG\rho}{2} + O\left(\frac{1}{c^2}\right).$$

Comparing this with (5.6), we conclude that $k = 8\pi$ and we can now write down *Einstein's equations* in their full glory:[1]

$$G_{\mu\nu} = \frac{8\pi G}{c^4} T_{\mu\nu}.$$

(5.14)

There are 10 equations here, as $T_{\mu\nu}$ is a symmetric 4×4 matrix. $R_{00} = \frac{8\pi G\rho}{c^4}\left(T_{00} - \frac{1}{2}g_{00}T\right)$ is the relativistic analogue of Poisson's equation (5.6); the other nine are new and describe the dynamics of the gravitational field, such as gravitational waves.

Recalling our first abortive attempt (5.3), we could add an extra term involving the metric $g_{\mu\nu}$ directly to the left-hand side and get a slightly more general equation:

$$G^{\mu\nu} + \Lambda g^{\mu\nu} = \frac{8\pi G}{c^4} T^{\mu\nu},$$

(5.15)

where Λ is a constant with dimensions of m^{-2}. This also satisfies all our assumptions and will be important when we come to apply Einstein's equation to cosmology.

In empty space, when there is no matter around and when $\Lambda = 0$, $T_{\mu\nu} = 0$ and Einstein's equations (5.14) are simply

$$G_{\mu\nu} = 0 \qquad \Leftrightarrow \qquad R_{\mu\nu} = 0.$$

Since $R_{\mu\nu}$ is constructed from second derivatives of the metric, an obvious solution is the Minkowski metric,

$$ds^2 = -c^2 dt^2 + dx^2 + dy^2 + dz^2 = \eta_{\mu\nu} dx^\mu dx^\nu,$$

for which the derivatives, and second derivatives, of the constant coefficients $\eta_{\mu\nu}$ all vanish. This is as it should be; flat space-time is a solution when there is no matter around to cause curvature. As we shall see later, the converse is not true; $R_{\mu\nu} = 0$ does *not* imply that space-time is flat.

Note that we can bring Λ onto the right-hand side of (5.15) and interpret it as matter with energy-momentum tensor

$$T_\Lambda^{\mu\nu} = -\frac{\Lambda c^4}{8\pi G} g^{\mu\nu}.$$

This is completely equivalent to a relativistic fluid (5.2) with $P + \rho c^2 = 0$ and $P = -\frac{\Lambda c^4}{8\pi G}$. The condition $P + \rho c^2 = 0$ gives a fluid whose equation of state is that the enthalpy vanishes (see problem 5.4).

---

[1] An alternative method of calculating k, using the Einstein tensor directly rather than the Ricci tensor, is given in problem 5.2.

5.2 The Physical Meaning of Einstein's Equations

Einstein's equations are for gravity what Maxwell's equations are for electromagnetism: they are both coupled sets of partial differential equations. But their nature is very different. Maxwell's equations are differential equations for the electric and magnetic fields distilled from observations such as Gauss' law and the Biot–Savart law. Gauss' law states that the total flux of electric field through a closed surface is proportional to the amount of electric charge contained within the surface

$$\int_S \mathbf{E}.d\mathbf{S} = \frac{Q_e}{\epsilon_0}$$

which can be re-expressed as the differential equation

$$\nabla.\mathbf{E} = \frac{\rho_e}{\epsilon_0}$$

where ρ_e is the density of electric charge.

Einstein's equations (5.14) have been written down directly as differential equations for the metric $g_{\mu\nu}$; there is no simple integral form for them. Feynman has, however, given a simple interpretation of the 00 equation,

$$G_{00} = \frac{8\pi G}{c^2}\rho,$$

so,

$$G_{00} = R_{00} - \frac{1}{2}g_{00}R = \frac{8\pi G}{c^2}\rho.$$

We are free to use any coordinate system we wish and, at any given point, we can use inertial coordinates for which $g_{\mu\nu} = \eta_{\mu\nu}$ at that point. We cannot do this before calculating $G_{\mu\nu}$. But, at any one point, we can choose to use inertial coordinates after doing all the derivatives to get $G_{\mu\nu}$: according to the equivalence principle, inertial coordinates correspond to freely falling observers. In an inertial coordinate system at the chosen point, $g_{00} = -1$ and $g_{\alpha\beta} = \delta_{\alpha\beta}$, and the Ricci tensor has components

$$R^0{}_0 = R^{\alpha 0}{}_{\alpha 0}, \quad R^\alpha{}_\beta = R^{0\alpha}{}_{0\beta} + R^{\gamma\alpha}{}_{\gamma\beta}, \quad R^0{}_\alpha = R^{\gamma 0}{}_{\gamma\alpha}.$$

The Ricci scalar is

$$R = R^0{}_0 + R^\alpha{}_\alpha = R^{0\alpha}{}_{0\alpha} + (R^{0\alpha}{}_{\alpha 0} + R^{\gamma\alpha}{}_{\gamma\alpha}) = 2R^0{}_0 + R^{\alpha\beta}{}_{\alpha\beta} = 2R^0{}_0 + R^{(3)},$$

where

$$R_{(3)} = R^{\alpha\beta}{}_{\alpha\beta}$$

is the 3-dimensional Ricci scalar of space only. Thus,

$$G_{00} = -G^0{}_0 = -R^0{}_0 + \frac{1}{2}R = \frac{1}{2}R_{(3)}$$

and the 00 component of Einstein's equations reads

$$R_{(3)} = \frac{16\pi G}{c^2}\rho. \tag{5.16}$$

This is not the whole content of Einstein's equations; it is only one part, but it tells us that the 3-dimensional Ricci scalar vanishes at any point where the mass/energy density vanishes.

For an inertial observer at any point on their world line, the mass density ρ at that point generates a curvature of 3-dimensional space. Non-relativistically we have Poisson's equation (1.10) and

$$R_{(3)} = \frac{4}{c^2}\nabla^2\Phi + O\left(\frac{1}{c^4}\right).$$

Equation (5.16) is the relativistic version of Poisson's equation.

Problems

1) Using the algebraic and differential properties of $R^\mu{}_{\nu\rho\sigma}$ described in Appendix B, verify that

$$G^\mu{}_{\nu;\mu} = 0.$$

2) Show that the Einstein tensor arising from the line element

$$ds^2 = -\left(1 + \frac{2\Phi(x,y,z)}{c^2}\right)c^2dt^2 + \frac{1}{\left(1 + \frac{2\Phi(x,y,z)}{c^2}\right)}(dx^2 + dy^2 + dz^2)$$

has components

$$G_{00} = \frac{2}{c^2}\nabla^2\Phi + \frac{4\Phi\nabla^2\Phi}{c^4} - \frac{5\nabla\Phi.\nabla\Phi}{c^4}$$

$$G_{0\alpha} = G_{\alpha 0} = 0$$

$$G_{\alpha\beta} = \frac{1}{(c^2 + 2\Phi)^2}\left((\nabla\Phi.\nabla\Phi)\delta_{\alpha\beta} - 2\partial_\alpha\Phi\partial_\beta\Phi\right),$$

where $x^0 = ct$. Use this result to determine k in (5.5) from the non-relativistic limit.

Note: the calculation is rather tedious and can be done using a symbolic manipulation language such as Mathematica™ or Maple™. (See question 4 in Chapter 6.)

3) Decompose the energy-momentum tensor for a relativistic fluid in Equation (5.2) into time-like and space-like parts T^{00}, $T^{0\alpha}$, and $T^{\alpha\beta}$ and examine their non-relativistic expansion in inverse powers of c in Minkowski space-time, with

$$g^{\mu\nu} = \eta^{\mu\nu} = \begin{pmatrix} -1 & 0 & 0 & 0 \\ 0 & 1 & 0 & 0 \\ 0 & 0 & 1 & 0 \\ 0 & 0 & 0 & 1 \end{pmatrix}$$

in Cartesian coordinates with $x^0 = ct$. Give a physical interpretation of the various components.

4) In thermodynamics the internal energy $U(S,V)$ is an extensive quantity, depending on the entropy S and the volume V, with

$$T = \left(\frac{\partial U}{\partial S}\right)_V, \qquad P = -\left(\frac{\partial U}{\partial V}\right)_S.$$

The enthalpy of a fluid $H(S,P)$ is the Legendre transform of the internal energy with respect to the volume V

$$H = U + PV.$$

Identifying the thermal energy density $\frac{U}{V}$ with ρc^2, show that a fluid with zero enthalpy is equivalent to a cosmological constant.

5) Calculate the trace of the energy-momentum tensor $T^\mu{}_\mu$ for a relativistic fluid. What is the value for a gas of thermal photons (black-body radiation)?

6

Solutions of Einstein's Equations in Empty Space

6.1 Schwarzschild Metric

We are now in a position to derive the Schwarzschild line element (2.31) from (5.14). In empty space, when there is simply no matter around, we have the Minkowski metric. Using 3-dimensional spherical polar coordinates, this is

$$ds^2 = -c^2 dt^2 + dr^2 + r^2 (d\theta^2 + \sin^2 \theta d\phi^2).$$

Now suppose we put a point mass M at the origin, or more generally a spherically symmetric mass distribution centred on the origin and containing a total mass M inside a radius r_0. Outside of r_0 there is no matter, space-time is empty, and Einstein's equations require $G_{\mu\nu} = R_{\mu\nu} = 0$ for $r > r_0$. Let us look for a time-independent, spherically symmetric solution of Einstein's equations,

$$G_{\mu\nu} = 0,$$

in the region $r > r_0$. From the symmetry of the problem, the most general form of the line element will be

$$ds^2 = -f^2(r)c^2 dt^2 + g^2(r)dr^2 + h^2(r)(d\theta^2 + \sin^2 \theta d\phi^2).$$

If we insist on looking for a static spherically symmetric solution, then the three functions $f(r)$, $g(r)$, and $h(r)$ must be independent of t, θ, and ϕ. We can in fact eliminate $h(r)$ immediately: define a new radial coordinate,

$$\tilde{r} = h(r),$$

then let

$$\tilde{g}(\tilde{r}) = g(r) \left(\frac{dr}{d\tilde{r}} \right)$$

126

and

$$\tilde{f}(\tilde{r}) = f(r),$$

so

$$ds^2 = -\tilde{f}^2(\tilde{r})c^2dt^2 + \tilde{g}^2(\tilde{r})d\tilde{r}^2 + \tilde{r}^2(d\theta^2 + \sin^2\theta d\phi^2).$$

All we have done is change variables; we might as well drop the tilde and use

$$ds^2 = -f^2(r)c^2dt^2 + g^2(r)dr^2 + r^2(d\theta^2 + \sin^2\theta d\phi^2).$$

We have not lost any generality in doing this.

We can calculate the Riemann tensor and the Ricci tensor generated by this metric, in terms of the two unknown functions $f(r)$ and $g(r)$ and their first and second derivatives. Details are given in Appendix D but for Einstein's equations all we need are the components of the Ricci tensor which are, with $x^0 = ct$,

$$R_{00} = \frac{1}{r^2}\frac{f}{g}\left(\frac{r^2 f'}{g}\right)'$$

$$R_{rr} = -\frac{1}{r^2}\frac{g}{f}\left(\frac{r^2 f'}{g}\right)' + \frac{2}{r}\frac{(fg)'}{fg}$$

$$R_{\theta\theta} = -\frac{r}{fg}\left(\frac{f}{g}\right)' + \frac{(g^2 - 1)}{g^2} \qquad (6.1)$$

$$R_{\phi\phi} = \sin^2\theta\, R_{\theta\theta}.$$

In empty space, with $\Lambda = 0$, outside the region containing the matter in a star, Einstein's equations demand that all the components in (6.1) should vanish (taking the trace of $G_{\mu\nu} = 0$ implies that $R = 0$, so $R_{\mu\nu} = 0$ too). $R_{\theta\theta} = 0$ and $R_{\phi\phi} = 0$ are, of course, the same equations, but it still seems that the system is overdetermined; there are three differential equations for two unknown functions. These are not really independent, though; they are related by the Bianchi identity (B.27), and there is a solution. Let's work through them carefully.

1) $R_{00} = 0 \Rightarrow \frac{r^2 f'}{g} = A$, where A is a constant with dimensions of length.

2) Using $R_{00} = 0$ in $R_{rr} = 0 \Rightarrow fg = B \Rightarrow g = \frac{B}{f}$, where B is a dimensionless constant, so

$$ff' = \frac{AB}{r^2} \quad \Rightarrow \quad f^2 = C - \frac{2AB}{r},$$

where C is a third constant.

3) Lastly, using $g = \frac{B}{f}$ in $R_{\theta\theta} = 0$ gives

$$-r(f^2)' + B^2 - f^2 = 0 \quad \Rightarrow \quad B^2 = C.$$

The unique solution is therefore remarkably simple:

$$f^2 = B^2 \left(1 - \frac{2A}{Br} \right), \qquad g^2 = \left(1 - \frac{2A}{Br} \right)^{-1}. \qquad (6.2)$$

A and B are two integration constants that necessarily appear because Einstein's equations involve derivatives of the metric. B can be eliminated by rescaling $t \to \frac{t}{B}$. This is essentially choosing different units for measuring time and we are free to rescale t to set $B = 1$.

We have proven that, up to a re-scaling of t, the unique spherically symmetric metric that satisfies Einstein's equations in empty space, $T_{\mu\nu} = 0$, is

$$ds^2 = -c^2 \left(1 - \frac{2A}{r} \right) dt^2 + \frac{dr^2}{\left(1 - \frac{2A}{r} \right)} + r^2 d\theta^2 + r^2 \sin^2\theta d\phi^2.$$

We saw in §3.7 that, in the $c \to \infty$ limit, this metric reproduces Newtonian gravity and Keplerian orbits around a mass M when $A = \frac{GM}{c^2}$. This is the justification for the Schwarzschild metric used in that section; it is the unique static metric that is spherically symmetric about a point and satisfies Einstein's equations in a region of space-time that contains no matter.

The fact that $R_{\mu\nu} = 0$ for the Schwarzschild metric does not imply that the space-time is flat. It was shown on page 114 that in general the Riemann tensor in four dimensions has 20 independent components. The Ricci tensor is a symmetric 4×4 matrix, so it has 10 components in four dimensions. Thus, Einstein's equations in flat space-time only fix 10 of the 20 components of the Riemann tensor; all of the curvature of the Schwarzschild metric is in those other 10 components, some of which are non-zero (see Appendix D).

6.2 Gravitational Waves

Despite the remarkable difference in structure between Einstein's dynamical theory of gravity, encoded in the field equations (5.14) and the dynamical equations governing electromagnetic forces, Maxwell's equations, they do share some features. For static fields the similarity of Newton's universal law of gravitation and Coulomb's law has already been remarked upon, and another feature that the corresponding dynamical equations have in common is the existence of wave-like solutions, gravitational waves in Einstein's case.

Wave-like solutions of his equations were found by Einstein himself in 1916, but there followed a long controversy as to whether or not they were real waves, capable of transporting energy, or were just a mathematical artefact of a coordinate transformation – Einstein himself changed his mind more than once on this point. It was finally settled in the 1960s by Hermann Bondi and co-workers in favour of the existence of gravitational waves, and experimentalists began to look for them. Again, there was controversy, and an early claim of detection in 1969 by Joseph Weber eventually proved to be false. Indirect evidence for gravitational waves later came from binary pulsar observations – two very massive compact objects orbiting around each other emit gravitational waves, thus losing energy, and the orbit decays, slowly but enough to be observed over a number of years. Russell Hulse and Joseph Taylor discovered a binary pulsar in 1974 and subsequently determined the orbital characteristics of the two pulsars orbiting around each other to an accuracy that allowed them to deduce that the orbit was decaying at precisely the rate predicted by general relativity, presumably due to energy loss through gravitational radiation, publishing their results in 1979. They received the Nobel Prize in Physics in 1993. Gravitational waves were finally detected directly in 2016 by a group in the United States, pioneered by the Scottish physicist Ron Drever. The 2017 Nobel Prize in Physics was awarded to Rainer Weiss, Barry Barish, and Kip Thorne for their discovery. (Ron Drever sadly died six months before the prize was awarded.)

Deriving the existence of gravitational waves from Einstein's field equations is more involved than deriving the existence of electromagnetic waves from Maxwell's equations, not only because Einstein's equations constitute a set of 10 coupled non-linear, second-order, partial differential equations for the 10 functions $g_{\mu\nu}$ (Maxwell's equations are a set of coupled, first-order, linear differential equations for six functions and therefore much easier to deal with), but also because of some subtleties associated with general coordinate invariance.

To show that there are wave-like solutions of Einstein's equations, we shall look at perturbations of Minkowski space-time. Consider a line element of the form

$$ds^2 = \eta_{\mu\nu}dx^\mu dx^\nu + h_{\mu\nu}dx^\mu dx^\nu$$

and

$$g_{\mu\nu} = \eta_{\mu\nu} + h_{\mu\nu}. \tag{6.3}$$

where $h_{\mu\nu}$ is small. The first subtlety is that we cannot apply the word 'small' to a quantity with dimensions. A kilometre is 1,000 when measured in metres, but it is 10^{-16} when measured in light-years. We had better make sure that all the components $h_{\mu\nu}$ are dimensionless. This is easily done; if x^μ are Cartesian coordinates for the Minkowski metric

$$\eta_{\mu\nu} = \begin{pmatrix} -1 & 0 & 0 & 0 \\ 0 & 1 & 0 & 0 \\ 0 & 0 & 1 & 0 \\ 0 & 0 & 0 & 1 \end{pmatrix},$$

then the coordinates all have dimensions of length, and the metric components are dimensionless and 'small' means each $h_{\mu\nu} \ll 1$.

In a region of space-time where there is no matter, $T_{\mu\nu} = 0$ and, with $\Lambda = 0$, Einstein's equations require $G_{\mu\nu} = 0$, and hence $R_{\mu\nu} = 0$. This does not mean that the metric must be flat; the Riemann tensor can be non-zero even if the Ricci tensor vanishes. Since $h_{\mu\nu} \ll 1$ by assumption, we can ignore terms quadratic in $h_{\mu\nu}$ or higher, and keep only terms linear in $h_{\mu\nu}$. This makes the calculation fairly straightforward, and in this approximation the components of the Riemann tensor are

$$R_{\mu\nu\rho\lambda} = \frac{1}{2}\left(\partial_\nu\partial_\rho h_{\mu\lambda} + \partial_\mu\partial_\lambda h_{\nu\rho} - \partial_\mu\partial_\rho h_{\nu\lambda} - \partial_\nu\partial_\lambda h_{\mu\rho}\right). \tag{6.4}$$

The Ricci tensor is then

$$R_{\mu\nu} = \frac{1}{2}\left(\partial_\mu\partial_\rho h^\rho{}_\nu + \partial_\nu\partial_\rho h_\mu{}^\rho - \Box h_{\mu\nu} - \partial_\mu\partial_\nu h\right),$$

where $\Box = \eta^{\mu\nu}\partial_\mu\partial_\nu = -\frac{1}{c^2}\partial_t^2 + \partial_x^2 + \partial_y^2 + \partial_z^2$ is the wave operator,[1] and $h = h^\rho{}_\rho$ the trace of $h_{\mu\nu}$.

The next subtlety is that, even when $h_{\mu\nu}$ are non-trivial functions, $R_{\mu\nu\rho\sigma}$ might still be zero, which means that (6.3) is the flat Minkowski metric in disguise. For example, if we change coordinates to

$$x^{\mu'} = x^\mu + \epsilon^\mu(x), \tag{6.5}$$

---

[1] Often referred to as the *d'Alembertian.*

where $\epsilon^\mu(x)$ are four differentiable functions with $\epsilon^\mu \ll 1$, then, to first order in ϵ_μ,

$$\eta_{\mu'\nu'} = \eta_{\mu\nu} + \partial_\mu\epsilon_\nu + \partial_\nu\epsilon_\mu \tag{6.6}$$

is still the flat metric, though not expressed in Cartesian coordinates, so $R_{\mu'\nu'\rho'\lambda'} = 0$ for the metric (6.6) – you should check this. Now here's the clever part – we can use this apparent complication to our advantage and simplify the formulae somewhat. Start with (6.3) and change coordinates to (6.5). Then, ignoring terms of order $\epsilon_\mu\epsilon_\nu$,

$$h_{\mu'\nu'} = h_{\mu\nu} + \partial_\mu\epsilon_\nu + \partial_\mu\epsilon_\nu.$$

It is then not too hard to show from (6.5), using the chain rule, that $\partial_{\mu'} = \partial_\mu - (\partial_\mu\epsilon^\lambda)\partial_\lambda$ and

$$\partial_{\nu'}h^{\nu'}{}_{\mu'} - \frac{1}{2}\partial_{\mu'}h^{\nu'}{}_{\nu'} = \partial_\nu h^\nu{}_\mu - \frac{1}{2}\partial_\mu h^\nu{}_\nu + \Box\epsilon_\mu.$$

So, if the four functions ϵ_μ are chosen to satisfy the linear inhomogeneous differential equation

$$\Box\epsilon_\mu = -\partial_\nu h^\nu{}_\mu + \frac{1}{2}\partial_\mu h^\nu{}_\nu, \tag{6.7}$$

then

$$\partial_{\nu'}h^{\nu'}{}_{\mu'} - \frac{1}{2}\partial_{\mu'}h^{\nu'}{}_{\nu'} = 0. \tag{6.8}$$

For any $h_{\mu\nu}$ it is always possible to find four functions ϵ_μ that satisfy (6.7); we just need to satisfy a linear inhomogeneous second-order differential equation for each ϵ_μ, and we know from the general theory of linear differential equations that this is always possible. Now if we express the Ricci tensor in the primed coordinates, we find

$$R_{\mu'\nu'} = -\frac{1}{2}\Box h_{\mu'\nu'},$$

and Einstein's equations in empty space $R_{\mu'\nu'} = 0$ imply

$$\Box h_{\mu'\nu'} = 0.$$

We can restrict $h_{\mu'\nu'}$ even further, though. We can make a further coordinate change from $x^{\mu'}$ to $x^{\tilde\mu} = x^{\mu'} + \tilde\epsilon^{\mu'}$ with $\tilde\epsilon^{\mu'}$ satisfying the homogeneous equation

$$\Box\tilde\epsilon^{\mu'} = 0$$

without disturbing (6.8). We now have

$$h_{\tilde\mu\tilde\nu} = h_{\mu'\nu'} + \partial_{\mu'}\tilde\epsilon_{\nu'} + \partial_{\nu'}\tilde\epsilon_{\mu'}$$

and we can choose $\tilde{\epsilon}_{\mu'}$ so that

$$h^{\tilde{\mu}}{}_{\tilde{\mu}} = h^{\mu'}{}_{\mu'} + 2\partial_{\mu'}\tilde{\epsilon}^{\mu'} = 0,$$

as this merely requires satisfying a single inhomogeneous first-order differential equation

$$\partial_{\mu'}\tilde{\epsilon}^{\mu'} = -\frac{1}{2}h^{\mu'}{}_{\mu'} \qquad (6.9)$$

relating the four functions $\tilde{\epsilon}^{\mu'}$, which is always possible.

The crucial point here is that there is a redundancy in the description, associated with our freedom to choose whatever coordinates we wish, and $h_{\mu\nu}$ are not all physically significant. Sometimes different $h_{\mu\nu}$ give the same geometry, and we can remove this redundancy by imposing some restrictions on $h_{\mu\nu}$. Equations (6.8) and (6.9) are examples of such possible restrictions, which are particularly convenient for discussing gravitational waves. They put 5 restrictions on the 10 functions $h_{\mu\nu}$, leaving only 5 functions free. Restricting the $h_{\mu\nu}$ in this way is called choosing a *gauge*, because another example of such a restriction is imposing a particular set of units on the coordinates (for example choosing to measure lengths in metres or light-years).[2]

We might as well drop the tildes on $x^{\tilde{\mu}}$ now and say that, if $h_{\mu\nu}$ satisfy the conditions

$$h^{\mu}{}_{\mu} = 0, \qquad \partial_{\nu}h^{\nu}{}_{\mu} = 0, \qquad (6.10)$$

then Einstein's equations imply that all 10 components $h_{\mu\nu}$ satisfy the wave equation

$$\Box h_{\mu\nu} = 0. \qquad (6.11)$$

The equations in (6.10) are gauge conditions; they are not essential and are not forced on us, but merely represent a choice of coordinates. Equations (6.11) are not gauge conditions; they are dynamical equations which physical gravitational fields must satisfy and they have wave-like solutions.

Again, Einstein's equations $R_{\mu\nu} = 0$ only fix 10 of the 20 components in the Riemann tensor; all the information about gravitational waves is contained in the other 10 that are still free. It is amusing to note that in three dimensions, the Riemann tensor only has six components – the same number as the Ricci tensor. There is no freedom in

---

[2] Equation (6.10) is sometimes called the *de Donder* gauge after one of the people to realise its utility in discussing gravitation waves, but it also has other names in the literature.

the Riemann tensor if the Ricci tensor is set to zero. There would be no
gravitational waves in a 3-dimensional space-time; gravitational waves
are fundamentally a 4-dimensional phenomenon.

We shall look for oscillating solutions of (6.11) of the form[3]

$$h_{\mu\nu} = P_{\mu\nu}e^{ik.x}, \tag{6.12}$$

where $P_{\mu\nu}$ is a constant real symmetric matrix and $k.x = k_\lambda x^\lambda$. The
gauge conditions (6.10) are satisfied if $P^\mu{}_\mu = 0$ and

$$k_\mu P^\mu{}_\nu = 0.$$

Einstein equations (6.11) will be satisfied if

$$k^2 = k_\mu k^\mu = 0.$$

A wave travelling in the x-direction is described by $k^\mu = \left(\frac{\omega}{c}, k, 0, 0\right)$ with

$$\frac{\omega^2}{c^2} - k^2 = 0$$

and

$$P_{\mu\nu} = \begin{pmatrix} 0 & 0 & 0 & 0 \\ 0 & 0 & 0 & 0 \\ 0 & 0 & P_+ & P_\times \\ 0 & 0 & P_\times & -P_+ \end{pmatrix}$$

with P_+ and $P_\times$ two real numbers. (The reason for the notation will
become clear soon.) $P_{\mu\nu}$ is called the *polarisation tensor*;[4] it describes
the possible polarisations of the gravitational wave. There are two
independent polarisations; the magnitude of each is given by P_+ and
$P_\times$.

To understand the structure of the gravitational wave, fix $x = 0$ and
consider the part of $h_{\mu\nu}$ in the y-z plane – the real part of (6.12) gives

$$\begin{pmatrix} h_{yy} & h_{yz} \\ h_{zy} & h_{zz} \end{pmatrix} = \begin{pmatrix} P_+ \cos(\omega t) & P_\times \cos(\omega t) \\ P_\times \cos(\omega t) & -P_+ \cos(\omega t) \end{pmatrix}.$$

Consider the two polarisations P_+ and $P_\times$ separately.

---

[3] Of course, $h_{\mu\nu}$ should be real functions for (6.3) to give an acceptable line element
and the geometry to be well defined. But everything we do will be linear in $h_{\mu\nu}$,
so we can use this complex notation and take the physical $h_{\mu\nu}$ to be just the real
part of (6.12).

[4] Not to be confused with the pressure tensor, which is a different beast.

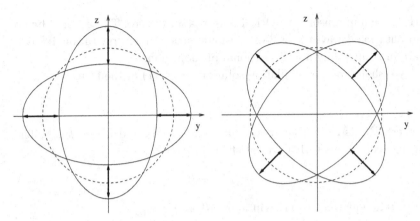

Figure 6.1 **Two polarisations of gravitational waves.** A circle in the *y-z* plane distorts into an ellipse as a gravitational wave moving in the *x*-direction passes. For P_+ the ellipse oscillates between having its long axis in the *y*-direction and in the *z*-direction. For $P_\times$ it oscillates between having the long axis in the direction $y = z$ and in the direction $y = -z$. A more general gravitational wave moving in the *x*-direction is a linear combination of these.

- For P_+ the line element in the *y-z* plane is

$$ds^2 = \left(1 + P_+ \cos(\omega t)\right)dy^2 + \left(1 - P_+ \cos(\omega t)\right)dz^2.$$

For $P_+ \ll 1$, a circle centred on the origin in the *y-z* plane at $t = 0$ will distort into an ellipse as the wave passes, oscillating 180° out of phase in the y and z directions (N–S and E–W compass directions) as shown in Figure 6.1. This circle can be thought of as a real physical object, like a loop of string or beads.

- For $P_\times$ the line element in the *y-z* plane is

$$ds^2 = dy^2 + dz^2 + 2P_\times \cos(\omega t)dy\,dz.$$

For $P_\times \ll 1$, a circle centred on the origin in the *y-z* plane at $t = 0$ will distort into an ellipse as the wave passes, oscillating 180° out of phase in the NW–SE and NE–SW directions.

These two polarisations are shown in Figure 6.1. The left-hand figure was used in §3.8 in the discussion of the gravitational wave geometry, and this section justifies the use of (3.60), with $P = -P_+$, as an approximate wave-like solution of Einstein's equations in empty space.

The propagation of gravitational waves through a vacuum has been described here, but the really interesting aspect is their production from accelerating matter, such as two neutron stars or two black holes orbiting one another. This is a very complex question, and any realistic analysis involves solving Einstein's equations numerically, which is beyond the scope of this book.

Problems

1) Calculate the Einstein tensor for a space-time with line element

$$ds^2 = -c^2 dt^2 + dr^2 + r^2 \big(d\theta^2 + \sin^2\theta (d\phi - \omega dt)^2\big),$$

with ω a constant angular velocity. How does the answer change if $\omega(t)$ becomes a non-trivial function of time?

2) Show that the Riemann tensor associated with the gravitational wave metric (6.3), with $h_{\mu\nu}$ given by (6.12), is non-zero, though the Einstein tensor vanishes.

3) Verify the connection coefficients and the components of the Riemann tensor for the line element

$$ds^2 = -f^2(r)c^2 dt^2 + g^2(r)dr^2 + r^2(d\theta^2 + \sin^2\theta d\phi^2) \qquad (6.13)$$

in Appendix D.

Evaluate the components of the Riemann tensor for the Schwarzschild line element, when $f(r) = \frac{1}{g(r)} = \sqrt{1 - \frac{2GM}{c^2 r}}$. Show that the Einstein tensor vanishes in this case.

4) A lot of the drudgery of calculating the Riemann and Einstein tensors can be done by symbolic manipulation programmes such as Mathematica™ or Maple™.

   i) If you have access to Mathematica™, download the **Einstein-Tensor** package and run the following script to determine the Einstein tensor arising from the line element (6.13):

   In[1] « Mathematica/EinsteinTensor.m
   In[2]:= x = {t,r,theta,phi}
   (* Spherical coordinates: *)
   In[3]:= (metric = DiagonalMatrix[{-(f[r])^2, (g[r])^2, r^2, (r Sin[theta])^2}]) //MatrixForm
   (* The line element: *)
   In[4]:= (Einstein = Simplify[EinsteinTensor[metric,x]])
   (*Calculate the Einstein tensor.*)

   ii) Alternatively, if you have access to Maple™, use the **Physics** package and run the following script:
   > with(Physics);
   > Setup(mathematicalnotation = true);

```
> ds2 := -f(r)^2 * dt^2 + g(r)^2 * dr^2 + r^2 * dtheta^2 + r^2 * sin(theta)^2 * dphi^2;
> Setup(coordinates = spherical, metric = ds2);
> CompactDisplay();
> G_11:= simplify(Einstein[1,1]);
> G_22:= simplify(Einstein[2,2]);
> G_33:= simplify(Einstein[3,3]);
> G_44:= simplify(Einstein[4,4]);
```

Show that the resulting Einstein tensor is diagonal with

$$G^t{}_t = \frac{\left(g - g^3 - 2rg'\right)}{r^2 g^3}$$

$$G^r{}_r = \frac{f - fg^2 + 2rf'}{r^2 f g^2}$$

$$G^\theta{}_\theta = G^\phi{}_\phi = \frac{\{-(f + rf')\,g' + g\,(f' + rf'')\}}{rfg^3}.$$

(It is recommended that the student still work through some examples, such as question 3, by hand – this gives an intuitive feeling for what is involved and helps lay the foundations for a better understanding of curvature and the Riemann tensor.)

5) The Schwarzschild metric describes space-time outside a static spherically symmetric body, such as a star (ignoring rotation). Suppose all the mass is contained in a radius R. Inside the body, still assuming spherical symmetry, there will be matter with density $\rho(r)$ sustained by a pressure $P(r)$. Define

$$m(r) = 4\pi \int_0^r \rho(r) r^2 dr, \tag{6.14}$$

with total mass

$$M = m(R).$$

For such a static spherically symmetric mass distribution, the energy-momentum tensor will be of the form

$$T^\mu{}_\nu(r) = \begin{pmatrix} -\rho(r)c^2 & 0 & 0 & \\ 0 & P(r) & 0 & 0 \\ 0 & 0 & P(r) & 0 \\ 0 & 0 & 0 & P(r) \end{pmatrix},$$

and the metric given by

$$ds^2 = -f^2(r)c^2 dt^2 + g^2(r)dr^2 + r^2(d\theta^2 + \sin^2\theta d\phi^2).$$

a) Using the results of question 4), show that the θ-θ and ϕ-ϕ Einstein equations imply that $f(r)$ and $g(r)$ are related by

$$\frac{(g(rf')' - (rf)'g')}{fg^3} = \frac{8\pi Gr P(r)}{c^4}.$$

b) Show that the 0-0 component of Einstein's equations is satisfied by

$$g(r) = \frac{1}{\sqrt{1 - \frac{2Gm(r)}{rc^2}}}.$$

c) Prove that the r-r Einstein equation requires

$$\frac{f'}{f} = \frac{\left(\frac{2Gm}{rc^2} + \frac{8\pi Gr^2 P}{c^4}\right)}{2r\left(1 - \frac{2Gm}{rc^2}\right)}.$$

d) Another equation for $\frac{f'}{f}$ can be obtained from the radial component of conservation of energy, $T^\mu{}_{r;\mu} = 0$. Use this to eliminate $\frac{f'}{f}$ and obtain the relation

$$\frac{dP}{dr} = -\frac{G\left(\rho + \frac{P}{c^2}\right)\left(m + \frac{4\pi r^3 P}{c^2}\right)}{r^2\left(1 - \frac{2Gm}{rc^2}\right)}.$$

This is known as the Tolman–Oppenheimer–Volkoff (TOV) equation; in the non-relativistic limit $c \to \infty$ it reduces to

$$\frac{dP}{dr} = -\frac{Gm(r)\rho(r)}{r^2}$$

which is the Newtonian equation for stellar equilibrium – for a star to be stable, the tendency to collapse under the gravitational force due to the mass enclosed by the radius r must be balanced by a pressure gradient.

e) The TOV equation is a differential equation involving two unknown functions, $\rho(r)$ and $P(r)$, ($m(r)$ and $\rho(r)$ are related by (6.14)). To solve it, we need a relation between ρ and P, the equation of state for the material inside the star. In general, the equation of state involves the temperature, but a very simple model is just to assume that $\rho = \rho_0$ is constant.[5]

[5] This may not be a bad approximation in the core of a star, but it cannot be correct all the way to the surface.

Solve the TOV equation when $\rho = \rho_0$ is constant, with the boundary condition that P vanishes at the surface of the star, $r = R$.

i) Show that

$$P = \rho_0 c^2 \left(\frac{\sqrt{1 - \frac{r^2 r_S}{R^3}} - \sqrt{1 - \frac{r_S}{R}}}{3\sqrt{1 - \frac{r_S}{R}} - \sqrt{1 - \frac{r^2 r_S}{R^3}}} \right)$$

where $M = \frac{4\pi}{3}\rho_0 R^3$ is the total mass of the star and $r_S = \frac{2GM}{c^2}$ is the Schwarzschild radius.

ii) Find $f(r)$ and $g(r)$ for constant ρ.

6) Consider the line element

$$ds^2 = -\left(1 - \frac{2GM}{c^2 r} - \frac{r^2}{L^2}\right)c^2 dt^2 + \frac{dr^2}{\left(1 - \frac{2GM}{c^2 r} - \frac{r^2}{L^2}\right)}$$
$$+ r^2(d\theta^2 + \sin^2\theta d\phi^2).$$

For $r \ll L$, this is the Schwarzschild space-time and, as such, is relevant for describing space-time outside a spherically symmetric, non-rotating star at the origin. Show that it is a solution of Einstein's equations in empty space with a cosmological constant $\Lambda = \frac{3}{L^2}$. As such, it may be viewed as describing the space-time outside a spherically symmetric, non-rotating star embedded in de Sitter space-time.

7

Cosmology and the Big Bang

On very large length scales ($>$ 100 Mpc $\approx$ 3×10^8 light years), the distribution of galaxies in the sky appears to be statistically the same in every direction (it is *isotropic*). If we assume that all points in space are equivalent, that is, we are not at a special point (called the *Copernican Principle*, because it resonates with Copernicus' realisation that our Earth is not the centre of the Solar System), then the distribution must also be isotropic about all other points. This actually implies that distribution of galaxies is *homogeneous* on very large scales, that is, it is uniform in space. We conclude that the mass density in galaxies ρ should be uniform and independent of position on large enough length scales, though it is not independent of time, as the Universe is not static.

Further evidence for an isotropic universe comes from the Cosmic Microwave Background (CMB). It was discovered in 1964 that intergalactic space contains thermal radiation; it is glowing at a temperature of 2.7 K. We shall see later that this can be interpreted as a smoking gun of the Big Bang. Space is filled with thermal radiation, which is a relic from the Big Bang 13.8 billion years ago. After allowance is made for the motion of our Solar System and contamination from hot matter from the Galactic plane, the temperature is extremely uniform, the largest deviations being only a few parts in 10^5 of the average (Figure 7.2). In fact, the spectrum of the CMB is the best black-body spectrum ever measured, as shown in Figure 7.3.

At the present day, the microwave background does not affect the dynamics of the Universe significantly; matter in and between galaxies dominates the energy density (though we shall see that the microwave background was very important in the early Universe). In principle, the average distribution of matter in galaxies, when smoothed out over scales

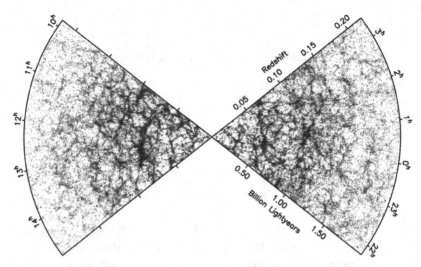

Figure 7.1 **Distribution of galaxies.** The distribution of galaxies over a region of space, out to a red-shift $z = 0.2$. Two thin slices spanning about 75° of the sky in opposite directions are depicted; each dot represents a galaxy. On large enough length scales, the distribution looks homogeneous and there is no preferred direction.
Image courtesy of the 2dFGRS Team.

significantly larger than intergalactic distances, could be very different from the distribution of the energy density of thermal radiation in the CMB, but this does not seem to be the case observationally. We have a very successful model of the early evolution of the Universe, to be described in this chapter, that predicts that these energy density distributions should be very similar; indeed, fluctuations in the distribution of the galaxies should largely trace fluctuations in the CMB temperature. It is a very reasonable assumption that, at least to a first approximation, the distribution of matter and energy in the Universe is isotropic and homogeneous on the largest accessible length scales.

On very large length scales (> 100 Mpc), we can model the mass distribution in the Universe as a smooth fluid with mass density ρ and pressure P. Think of the galaxies themselves as being like fluid particles with no internal structure, and view the fluid from a large enough length scale that the individual particles are not distinguishable; they just meld together to give the impression of a smooth fluid. Most galaxies are less than 0.1 Mpc in size, and on average galaxies are a few Mpc apart, so there could be a million galaxies in a region of space 100 Mpc across. As observers, we are attached to one particular fluid particle (our Galaxy)

Figure 7.2 **All sky map of the temperature of the Cosmic Micro-wave Background.** A plot of the temperature of the CMB over the whole sky. It is extremely smooth (upper image); no deviations from the value 2.725 K are visible, indicating a very high degree of isotropy. When the deviations from the average temperature are magnified, tiny differences in temperature in different directions do show up (lower image). This is indicative of small inhomogeneities at the order of a few parts in 10^5. ©ESA/Planck Collaboration.

and we shall adopt a Lagrangian (as opposed to a Eulerian) mechanics point of view, in which coordinates are attached to the particles themselves (called the comoving coordinate picture) rather than to abstract points in space. From our local perspective then, the fluid is at rest and so has zero momentum and $p^\alpha = 0$ in the energy-momentum tensor (5.1).

Furthermore, the observed isotropy constrains the 3×3 pressure matrix, $P^{\alpha\beta}$, to be proportional to the identity matrix. (This is in keeping with our picture of a non-viscous fluid; non-viscous fluids cannot

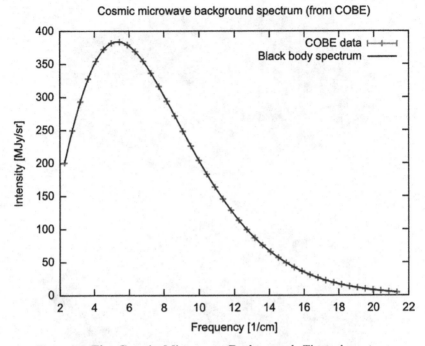

Figure 7.3 **The Cosmic Microwave Background.** Thermal spectrum of the CMB (measured by the Far Infrared Absolute Spectrophotometer (FIRAS), NASA/COBE science team). The intensity in radiation as a function of frequency. The curve is a perfect example of a black-body spectrum at 2.725 K. (The error bars are smaller than the thickness of the curve.)

sustain tangential pressure, so the off-diagonal pressures in the energy-momentum tensor are zero, and the diagonal entries must be equal for isotropy.) For a fluid at rest in flat (Minkowski) space-time, using Cartesian coordinates, the energy-momentum tensor has the form (C.11)

$$T^{\mu\nu} = \begin{pmatrix} \rho c^2 & 0 & 0 & 0 \\ 0 & P & 0 & 0 \\ 0 & 0 & P & 0 \\ 0 & 0 & 0 & P \end{pmatrix}, \tag{7.1}$$

where P is the pressure.[1] More generally in a curved space-time (or using curvilinear coordinates in Minkowski space-time), it would be of the form (C.14), where U^μ are the components of the 4-velocity of the fluid.

---

[1] Not to be confused with the polarisations, P_+ and $P_\times$ in Chapter 5.

For ordinary matter made up of stars in galaxies, we might expect $P = 0$, but most galaxies do not form in isolation; they form in clusters, and evidence from X-rays indicates that there is a lot of mass in ionised gas clouds between galaxies in clusters, even more mass than there is in stars, so it would be more general to include P. Actually it is still the case that the pressure contribution to $T_{\mu\nu}$ from this intergalactic ionised matter is negligible in terms of cosmological dynamics, and ρ dominates over P for ordinary matter. However, thermal radiation (a photon gas) was important in the dynamics of the early Universe, and for thermal radiation, $P = \frac{1}{3}\epsilon$, where ϵ is the energy density in the radiation (equivalent to ρc^2 for ordinary matter). We shall therefore retain P in the general formalism.

The assumption that the Universe is homogeneous and isotropic about every point in space constrains the form of the metric. If 3-dimensional space is flat, it can only be of the form

$$ds^2 = -b^2(t)c^2dt^2 + a^2(t)(dx^2 + dy^2 + dz^2),$$

where $b(t)$ and $a(t)$ are functions of time. Actually, b is redundant; we can always define a new time parameter, $\tilde{t}$, by

$$\tilde{t} = \int^t b(t)dt,$$

in terms of which

$$ds^2 = -c^2d\tilde{t}^2 + a^2(\tilde{t})d\mathbf{x}.d\mathbf{x}.$$

Now drop the tilde, and we have:

$$ds^2 = -c^2dt^2 + a^2(t)d\mathbf{x}.d\mathbf{x}. \tag{7.2}$$

This is the *Robertson–Walker* metric.

With $a(t)$ a function of t, we should expect $\rho(t)$ and $P(t)$ in (7.1) to depend on t as well. Referring to (C.14) with $U^\mu = (c,0,0,0)$ and $g^{\mu\nu}$ given by

$$g^{\mu\nu} = \begin{pmatrix} -1 & 0 & 0 & 0 \\ 0 & \frac{1}{a^2(t)} & 0 & 0 \\ 0 & 0 & \frac{1}{a^2(t)} & 0 \\ 0 & 0 & 0 & \frac{1}{a^2(t)} \end{pmatrix}$$

(again using $x^0 = ct$), we write

$$T^{\mu\nu} = \begin{pmatrix} \rho(t)c^2 & 0 & 0 & 0 \\ 0 & \frac{P(t)}{a^2(t)} & 0 & 0 \\ 0 & 0 & \frac{P(t)}{a^2(t)} & 0 \\ 0 & 0 & 0 & \frac{P(t)}{a^2(t)} \end{pmatrix}, \tag{7.3}$$

$$T_{\mu\nu} = \begin{pmatrix} \rho(t)c^2 & 0 & 0 & 0 \\ 0 & a^2(t)P(t) & 0 & 0 \\ 0 & 0 & a^2(t)P(t) & 0 \\ 0 & 0 & 0 & a^2(t)P(t) \end{pmatrix} \tag{7.4}$$

for a fluid at rest.

The functional form of the cosmic scale factor $a(t)$ will be determined by solving Einstein's equations, which reduce to a second-order differential equation for $a(t)$ in terms of $\rho(t)$ and $P(t)$. But we can immediately say something using conservation of energy and momentum, as expressed by the equation

$$T^{\mu\nu}{}_{;\mu} = \partial_\mu T^{\mu\nu} + \Gamma^\mu_{\rho\mu} T^{\rho\nu} + \Gamma^\nu_{\rho\mu} T^{\mu\rho} = 0 \tag{7.5}$$

(see Appendix C). The connection coefficients for the line element (7.2) are given in Appendix E, and using these, the four conservation equations (7.5) are

$$\dot{\rho}c^2 + 3\rho c^2 \frac{\dot{a}}{a} + 3P\frac{\dot{a}}{a} = 0 \quad \Rightarrow \quad \frac{\partial}{\partial t}(\rho a^3) = -\frac{P}{c^2}\frac{\partial}{\partial t}(a^3), \tag{7.6}$$

$$\frac{\partial P}{\partial x^\alpha} = 0, \qquad \alpha = 1, 2, 3.$$

The equation $\partial_\alpha P = 0$ is built in to the construction; we have assumed that space is homogeneous, so $P(t)$ in (7.1) is assumed to be independent of position, as is the line element (7.2). Indeed, we could have used $\frac{d}{dt}$ rather than $\frac{\partial}{\partial t}$ in (7.6). The first equation in (7.6) tells us that when the pressure is small, or in the non-relativistic limit when c is large, the total mass contained in a volume a^3, ρa^3, is constant (this is just conservation of mass). Relativistically, when $\dot{a} \neq 0$, the pressure modifies the law of conservation of mass, because the pressure contributes to the energy density.

More generally, the 3-dimensional Euclidean line element $ds_E^2 = d\mathbf{x}^2.d\mathbf{x}^2$ can be replaced by the line element of any 3-dimensional isotropic homogeneous space and, up to re-scalings, there are only three possibilities: flat Euclidean space, a space of uniform positive curvature (a 3-dimensional sphere), or a space of uniform negative curvature (a 3-dimensional version of the Lobachevsky plane).[2] These are generalisations of the 2-dimensional spaces described in §2.3 and §2.5: Equation (2.13) with the circle (line element $r^2 d\phi^2$) replaced by a 2-dimensional sphere (line element $r^2(d\theta^2 + \sin^2\theta d\phi^2)$). The only possible isotropic, homogeneous, 3-dimensional metrics are described by the line elements

$$ds^2 = a^2 \left(\frac{dr^2}{1 - Kr^2} + r^2(d\theta^2 + \sin^2\theta d\phi^2) \right),$$

with K a constant. So allowing for non-zero, but uniform, spatial curvature generalises Equation (7.2) to

$$ds^2 = -c^2 dt^2 + a^2(t) \left(\frac{dr^2}{1 - Kr^2} + r^2(d\theta^2 + \sin^2\theta d\phi^2) \right). \qquad (7.7)$$

For uniform curvature, it is always possible to re-scale the spatial coordinates so that $K = +1$, -1, or 0, depending on whether the curvature is positive, negative, or space is flat. The observational evidence, which we shall describe later, is that 3-dimensional space in our Universe is actually flat, but we shall keep K explicit for the moment. K then appears in the equations below, and we see that spatial curvature, were it present in any degree, would affect the dynamics of the Universe.

Observations also tell us that in order to model the dynamics of our Universe at the largest observable scales, it is necessary to include a cosmological constant Λ. So we use the full Einstein equation (5.15) with $T_{\mu\nu}$ given in (7.1) and the Einstein tensor $G_{\mu\nu}$ calculated in terms of the function $a(t)$ in the Robertson–Walker line element (7.2).

The Einstein tensor for the Robertson–Walker line element is calculated in Appendix E and given in Equation (E.6). Einstein's equations with Λ then reduce to just two differential equations: the G_{00} equation is

$$\frac{3}{a^2} \left(\dot{a}^2 + Kc^2 \right) = 8\pi G\rho + \Lambda c^2 \qquad (7.8)$$

and, because of our assumed spatial isotropy, all the other equations reduce to just one equation

$$-\frac{1}{a^2} \left(2\ddot{a}a + \dot{a}^2 + Kc^2 \right) = \frac{8\pi G}{c^2} P - \Lambda c^2. \qquad (7.9)$$

These two equations are known as the *Friedmann equations* and they govern the dynamics of the Universe on the largest scales.

---

[2] In a general dimension 'positive' and 'negative' curvature here refer to the sign of the Ricci scalar.

7.1 The Friedmann Equations

We now wish to solve the Friedmann equations (7.8) and (7.9) to find the functional form of $a(t)$ for physically reasonable density and pressures. The first Friedmann equation (7.8) can be written as

$$\left(\frac{\dot{a}}{a}\right)^2 = \frac{8\pi G\rho}{3} + \frac{\Lambda c^2}{3} - \frac{c^2 K}{a^2}. \tag{7.10}$$

Eliminating $\dot{a}^2 + Kc^2$ from (7.8) and (7.9) gives

$$-\frac{3\ddot{a}}{a} = 4\pi G\left(\rho + \frac{3P}{c^2}\right) - \frac{\Lambda}{c^2}. \tag{7.11}$$

The minus sign on the left-hand side of this equation has a simple physical interpretation when P and Λ are zero – positive ρ makes $\ddot{a}$ negative; gravity is always attractive, so matter causes the expansion to decelerate. However, if Λ were positive and large enough to dominate the density and pressure in (7.11), $\ddot{a}$ could become positive, and the expansion of the Universe would accelerate.

Alternatively, differentiating $a^2 \times$ (7.10) with respect to t eliminates K and using (7.11) to eliminate $\ddot{a}$ leads to

$$\dot{\rho}a^2 + 3\rho a\dot{a} + \frac{3P}{c^2}a\dot{a} = 0$$

$$\Rightarrow \quad \frac{d}{dt}\left(\rho a^3\right) = -\frac{3P}{c^2}a^2\dot{a}, \tag{7.12}$$

which is just the conservation equation we had before (7.6), so (7.11) tells us nothing new[3] ((7.6) is unchanged when $K = \pm 1$). We can focus on (7.10) and (7.12).

The combination $\frac{\dot{a}}{a}$ appears frequently from now on, and it is convenient to define the function

$$H(t) = \frac{\dot{a}}{a}. \tag{7.13}$$

As described in §2.6.1, the present-day value of H can be ascertained by measuring the red-shift of receding galaxies and using the red-shift-distance relation. Let t_0 be the present time and $H_0 = H(t_0)$ the present-

---

[3] It is no coincidence that Einstein's equations reproduce (7.6). We derived Einstein's equations by using the second Bianchi identity (B.27) to guide us – when $T_{\mu\nu}$ is related to $G_{\mu\nu}$ through Einstein's equations, (7.12) is essentially the Bianchi identity.

day value of H. The current best estimate from red-shift measurements was given in Equation (2.30); it is

$$H_0 = 73.2 \pm 1.3 \, \text{km s}^{-1} \, \text{Mpc}^{-1}. \tag{7.14}$$

H_0 is called the *Hubble constant*, because it is the constant of proportionality that appears in the red-shift-distance relation (2.29), but $H(t)$ is in general a function of time.

In terms of $H(t)$, the Friedmann equation (7.10) reads

$$\boxed{H^2 = \frac{8\pi G\rho}{3} + \frac{\Lambda c^2}{3} - \frac{Kc^2}{a^2}}, \tag{7.15}$$

and much of the rest of this chapter will be devoted to studying this equation.

In the Friedmann equation, Λ and K are constants, and more will be said about their values later. Since the Universe is expanding, we should expect ρ to be a function of time, but we can estimate its present-day value by counting galaxies. Observationally, counting the visible luminous galaxies[4] gives $\rho_{Luminous} \approx 3 \times 10^{-29}$ kg m^{-3} while (7.14) gives

$$\frac{3H_0^2}{8\pi G} = 1.0 \times 10^{-26} \, \text{kg m}^{-3}. \tag{7.16}$$

Clearly, the Friedmann equation (7.15) cannot be satisfied by the matter density in luminous galaxies alone; either there must be more matter or Λ and/or K is non-zero (or some combination of these).

From observations of the orbits of stars in galaxies around the galactic centres, we can use Kepler's laws to deduce how much matter there is in the galaxy to make the stars follow their observed trajectories, and there is clear evidence that almost all galaxies have more matter in their peripheral regions than is visible in luminous stars. Similarly, using observations of orbital dynamics of galaxies in galactic clusters, we can estimate the amount of matter in the cluster that is causing the gravitational attraction keeping the cluster together, and this gives $\rho \approx 10^{-27}$ kg m^{-3} (equivalent to about one proton per cubic metre); again, there is much more matter in the cluster than is visible in stars.

---

[4] On average, galaxies are about $10 \, \text{Mpc} \approx 3 \times 10^{23}$ m apart, and a typical galaxy contains about 10^{11} stars. A typical star, such as the Sun, has a mass of about 2×10^{30} kg, so an order of magnitude estimate is

$$\frac{(2 \times 10^{30}) \times 10^{11}}{(3 \times 10^{23})^3} \, \text{kg m}^{-3} \approx 10^{-29} \, \text{kg m}^{-3}.$$

Most of the matter in the Universe is not visible. The most accurate value we have to date comes from the dynamics of galaxies on cosmological scales, which yields $\rho = 2.5 \times 10^{-27}$ kg m^{-3}, more than is visible, but still not enough for (7.16) – either Λ or K must play a role. To make further progress, we need to solve the full Equation (7.15), and to do this we need to know more about the functional form of $\rho(t)$.

The density is related to the pressure via the conservation equation (7.12), so we have two differential equations, (7.12) and (7.15), for three unknown functions, $\rho(t)$, $P(t)$, and $a(t)$. We need a further relation to solve the system, and this comes in the form of a relation between ρ and P called the *equation of state*.

A simple example of an equation of state is the ideal gas law $PV = Nk_BT$ for N gas particles at a temperature T in a volume V (k_B is Boltzmann's constant). For a monatomic gas of particles with mass m, the density is $\rho = \frac{Nm}{V}$, so

$$P = \frac{\rho}{m} k_B T;$$

for a given temperature, the pressure is proportional to the density – this is the equation of state for a monatomic ideal gas. But the ideal gas law is not relevant to the description of galaxies – there is no short-range repulsion between galaxies the way there is between the particles in a normal gas. Indeed, non-relativistically, pressure plays no role, due to the factor of $\frac{1}{c^2}$ on the right-hand side of (7.12). To a good approximation, there is *no* pressure between the galaxies, so we can just set $P = 0$. An evocative term for a fluid with zero pressure is to call it a *dust*. We treat galaxies like particles of dust with mass but no pressure. So, our equation of state will be

$$P = 0.$$

We can now immediately solve (7.12); ρa^3 is constant, independent of time. This has a very natural interpretation: suppose $K = 0$ so that space is flat; then the total mass inside a solid ball of radius a is $\frac{4\pi}{3}\rho a^3$. For $K \neq 0$, the prefactor $\frac{4\pi}{3}$ is changed, but the interpretation is the same: when $P = 0$, (7.12) states that the total amount of mass in a comoving volume of space is constant; mass is being neither created nor destroyed. We can now forget (7.12) and focus on (7.15) with $\rho \propto \frac{1}{a^3}$.

Start by writing it as

$$H^2 = \frac{8\pi G\rho}{3} - \frac{c^2 K}{a^2} + \frac{\Lambda c^2}{3} \quad \Rightarrow \quad 1 = \frac{8\pi G\rho}{3H^2} - \frac{c^2 K}{H^2 a^2} + \frac{\Lambda c^2}{3H^2}.$$

Now define the following three constants using present-day values H_0 and $a_0 = a(t_0)$, where t_0 is the present day:

$$\Omega_M := \frac{8\pi G \rho(t_0)}{3 H_0^2}, \quad \Omega_K := -\frac{c^2 K}{H_0^2 a_0^2} \quad \text{and} \quad \Omega_\Lambda := \frac{\Lambda c^2}{3 H_0^2}.$$

Only two of these are independent, since

$$\Omega_M + \Omega_K + \Omega_\Lambda = 1.$$

These cosmological Ω's can be related to the coefficients in a Taylor expansion of $a(t)$ about the present day,

$$a(t) = a_0 + (t - t_0)\dot{a}_0 + \frac{1}{2}(t - t_0)^2 \ddot{a}_0 + \dots$$

$$= a_0 \left(1 + H_0(t - t_0) + \frac{1}{2}(t - t_0)^2 \frac{\ddot{a}_0}{a_0} + \dots \right),$$

where, in what is meant to be an obvious notation, $\dot{a}_0 = \dot{a}(t_0)$ and $\ddot{a}_0 = \ddot{a}(t_0)$. Define the *deceleration parameter*,[5] $q_0 = -\frac{1}{2}\frac{\ddot{a}_0}{H_0^2 a_0}$; then since

$$\dot{a}^2 = \frac{8\pi G \rho a^2}{3} - c^2 K + \frac{\Lambda c^2 a^2}{3},$$

and, from (7.12) with $P = 0$, $\rho \propto \frac{1}{a^3}$ so

$$\rho = \frac{3}{8\pi G} \frac{A}{a^3} \tag{7.17}$$

for some constant A. In terms of A, the Friedmann equation reads

$$\dot{a}^2 = \frac{A}{a} - c^2 K + \frac{\Lambda c^2 a^2}{3}$$

$$\Rightarrow \quad 2\dot{a}\ddot{a} = -\frac{A\dot{a}}{a^2} + \frac{2\Lambda c^2 a\dot{a}}{3}$$

$$\Rightarrow \quad \frac{\ddot{a}}{a} = -\frac{A}{2a^3} + \frac{\Lambda c^2}{3} = -\frac{4\pi G \rho}{3} + \frac{\Lambda c^2}{3}$$

$$\Rightarrow \quad \frac{\ddot{a}}{H^2 a} = -\frac{4\pi G \rho}{3 H^2} + \frac{\Lambda c^2}{3 H^2}.$$

Evaluating this at t_0 gives

$$2q_0 = \frac{1}{2}\Omega_M - \Omega_\Lambda.$$

---

[5] Why the minus sign? Historically it was usually assumed that $\Lambda = 0$ and the expectation was that the expansion of the Universe would be slowing down when a got large enough to ignore the K-term in (7.15). The expansion would then decelerate, due to the gravitational attraction of ρ, and q_0 would be positive. Wrong! Current observations indicate that Λ is positive and q_0 is negative, so the expansion is accelerating.

Thus, Ω_Λ and Ω_M can be directly related to the Taylor expansion of $a(t)$ in terms of H_0 and q_0.

What we actually measure is the red-shift z, but we can convert from t to z. From (2.27), with $t_2 = t_0$ and $t_1 = t$, and the definition $z = \frac{\nu_1}{\nu_2} - 1$ we have $\frac{a_0}{a(t)} = 1 + z$

$$\Rightarrow \quad dz = -\frac{a_0}{a^2}da \quad \Rightarrow \quad \frac{dz}{1+z} = -\frac{da}{a} \quad \Rightarrow \quad \frac{\dot{a}}{a} = -\frac{\dot{z}}{1+z}.$$

Now use this to write the Friedmann equation as

$$H(t)^2 = \left(\frac{\dot{a}}{a}\right)^2 = \frac{A}{a^3} - \frac{c^2 K}{a^2} + \frac{\Lambda c^2}{3}$$

$$= \Omega_M H_0^2 \left(\frac{a_0}{a}\right)^3 + \Omega_K H_0^2 \left(\frac{a_0}{a}\right)^2 + \Omega_\Lambda H_0^2$$

$$\Rightarrow \quad H(t)^2 = H_0^2 \left\{\Omega_M (1+z)^3 + \Omega_K (1+z)^2 + \Omega_\Lambda\right\} \tag{7.18}$$

or

$$\dot{z}^2 = H_0^2 (1+z)^2 \left\{(1+z)^3 \Omega_M + (1+z)^2 \Omega_K + \Omega_\Lambda\right\}. \tag{7.19}$$

This is a non-linear differential equation (no approximations) for $z(t)$ in terms of the four constants H_0, Ω_M, Ω_K, and Ω_Λ.

Referring back to the discussion on expanding space in §2.6, we invoke (2.25) with our Galaxy at $r = 0$ and a distant galaxy at r_A. As we have repeatedly tried to emphasise, r_A is just a coordinate attached to galaxy A; it is not a physical distance. Actually, the definition of 'distance' in general relativity can be subtle. We might define the distance d_A to a galaxy A at the present day t_0 as

$$d_A = a_0 r_A.$$

This is one possible definition, motivated by putting $dt = 0$ in the line element. Imagine closing your eyes and visualising where all the galaxies are at exactly the same moment of time (this is not really physical, as there is no universal 'moment of time'; different observers would define time differently). Then imagine putting metre sticks together end to end to stretch to galaxy A. At any chosen fixed t, the total distance on the metre sticks would amount to $d_A(t) = a(t) r_A$. We might call d_A the 'metre stick' distance to galaxy A. (Though this method of measuring distances is not without practical difficulties.)

Another definition of distance would be to measure the time it takes a beam of light to travel from A to our Galaxy at $r = 0$ and multiply by the speed of light. The light, coming radially inwards, follows a trajectory with

$$cdt = -a(t)dr,$$

so if the light leaves A at time t and arrives here at time t_0, then, from (2.25),

$$\int_t^{t_0} \frac{dt}{a(t)} = -\frac{1}{c}\int_{r_A}^0 dr = \frac{r_A}{c}. \tag{7.20}$$

Then $s_A = (t_0 - t)c$ is a distance that we might call 'time of flight distance'. Time of flight distance and metre stick distance are not the same in general, though they can be identified for small z and are the same if $a(t)$ is a constant.

With this warning in mind, we shall press ahead using metre stick distance d_A. From (7.18) and (7.20),

$$d_A = a_0 r_A = -ca_0 \int_t^{t_0} \frac{dt}{a} = -c\int_t^{t_0}(1+z)dt = -c\int_t^{t_0}\frac{(1+z)dz}{\dot{z}},$$

finally giving

$$\boxed{d_A = \frac{c}{H_0}\int_0^z \frac{dz}{\sqrt{(1+z)^3\Omega_M + (1+z)^2\Omega_K + \Omega_\Lambda}}.} \tag{7.21}$$

This is an exact non-linear red-shift-distance relation for which the Hubble relation (2.29) is the linear approximation; deviations from linearity are only apparent for z near 1 and above, as shown in Figure 2.6.

For example,

$$d_A(z) = \frac{c}{H_0}z \tag{7.22}$$

if $\Omega_\Lambda = 1$, $\Omega_M = \Omega_K = 0$, while

$$d_A(z) = \frac{2c}{H_0}\left(1 - \frac{1}{\sqrt{1+z}}\right)$$

if $\Omega_M = 1$, $\Omega_\Lambda = \Omega_K = 0$.

Although the case (7.22) looks identical to (2.29), they have different interpretations since $d_A = a_0 r_A$ is not the same as $s_A = c(t_0 - t)$ (with $t_A = t$ and $t_B = t_0$ in (2.29)), though their difference is negligible at small $z \ll 1$. Evidence that $\Lambda > 0$ can be seen in Figure 2.6 near $z = 1$, where different curves are fitted for a number of different values of $(\Omega_M, \Omega_\Lambda)$. The best fit comes from collating data from a number of sources and gives

$$\Omega_M = 0.3089 \pm 0.0062 \begin{cases} 0.0487 \pm 0.0011 & \text{'ordinary matter'} \\ & \text{(neutrons, protons)} \\ 0.266 \pm 0.017 & \text{'Non-baryonic dark matter'} \\ & \text{(properties unknown)} \end{cases}$$

$\Omega_\Lambda = 0.6911 \pm 0.0062$ often called 'Dark Energy'.

$$1 = \Omega_M + \Omega_K + \Omega_\Lambda \;\Rightarrow\; \Omega_K = -0.000 \pm 0.024$$

$$\Omega_K = 0 \;\Rightarrow\; 1 = \Omega_M + \Omega_\Lambda \quad \Rightarrow \quad q_0 = \frac{1}{2}\left[\frac{1}{2}(1-\Omega_\Lambda) - \Omega_\Lambda\right]$$
$$= \frac{1}{4}(1 - 3\Omega_\Lambda) = -0.264 \pm 0.013,$$

so q_0 is negative, and the expansion rate of the Universe is accelerating.

A value of 0.31 for Ω_M means that the density in matter at the present time is

$$\rho = 0.31 \times \frac{3H_0^2}{8\pi G} = 3.1 \times 10^{-27}\text{kg m}^{-3}, \tag{7.23}$$

not far off the value of 2.5 that we get from galactic dynamics, but two orders of magnitude more than the amount of luminous matter that can be seen in stars ($\rho_{Luminous} \approx 3 \times 10^{-29}$ kg m^{-2}). We shall see later, when we discuss the formation of the chemical elements in the early Universe, that (7.23) cannot all be due to ordinary matter with which we are familiar, protons and neutrons (known as baryonic matter, *baryon* being the term used in particle physics for a class of particles that includes protons and neutrons). Only 4.5×10^{-28} kg m^{-3} (about 15 per cent of (7.23)) can be protons and neutrons; the other 85 per cent appears to be some new form of matter about which almost nothing is known – it has never been detected in a laboratory; we only infer its existence from cosmological observations. This is called 'non-baryonic dark matter'. Only about 1 per cent of the matter in our Universe is visible; 99 per cent is 'dark' and of that 99 per cent, only something like 15 per cent is baryonic (protons and neutrons) and 85 per cent is something completely new, non-baryonic dark matter. This last 85 per cent is one of the greatest mysteries of modern physics.

7.1.1 Analytic Solutions of the Friedmann Equation

With $\rho \propto \frac{1}{a^3}$ the Friedmann equation (7.15) is

$$\dot{a}^2 = \frac{A}{a} - c^2 K + \frac{\Lambda c^2 a^2}{3}. \tag{7.24}$$

This is a non-linear first-order differential equation for the unknown function $a(t)$, depending on three constants, Λ, A, and K or, equivalently, Ω_Λ, Ω_M, and Ω_K. Finding a unique solution requires setting a boundary condition. We need to avoid $a(t) < 0$ and, given that the Universe is expanding, a natural choice, if possible, is to set $a(0) = 0$, indicating that the cosmological scale factor was zero a finite time ago.

Solving this equation with all three of A, Λ, and K being non-zero is best done using numerical integration. However, even without trying to find analytic solutions, we can get a feeling for how things are going to go by analogy with 1-dimensional particle mechanics. Think of $a(t)$ as the position of a particle with unit mass moving on the positive real line under the influence of a potential:

$$\underbrace{\frac{1}{2}\dot{a}^2}_{\text{kinetic energy}} \underbrace{- \frac{1}{2}\left[\frac{\Lambda c^2 a^2}{3} + \frac{A}{a}\right]}_{\text{potential energy}} = \underbrace{-\frac{c^2 K}{2}}_{\text{total energy}}. \tag{7.25}$$

Then $V(a) = -\frac{A}{2a} - \frac{\Lambda c^2 a^2}{6}$ is like the potential energy per unit mass of a particle moving in one dimension, a combination of an inverse square $\frac{1}{a^2}$ attractive force (like Newtonian gravity) and a linear a repulsive force (like a repulsive harmonic oscillator). Positive K is negative energy, and negative K is positive energy.

Referring to Figure 7.4, the potential goes to minus infinity at $a = 0$ if $A > 0$, so $\dot{a}$ diverges there. Solutions like this start from a Big Bang a finite time ago. There is no contradiction with relativity here, though; nothing is travelling faster than light, $\dot{a}$ is not a physical velocity.

The behaviour of specific solutions depends on K and the sign of Λ, among other things. One can have fun playing with the various possibilities. For example, one amusing solution of (7.25) is obtained by setting $A = \Lambda = 0$ and $K = -1$; then the solution is a linear function:

$$a = ct.$$

Referring to (E.7), the Riemann tensor vanishes for $a = ct$ and $K = -1$, so this space-time is flat. This is a 4-dimensional version of the 2-dimensional Milne universe discussed in §2.6.2; it is flat space-time in an unusual coordinates system.

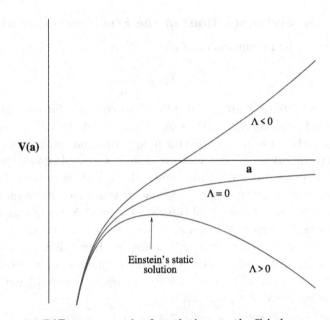

Figure 7.4 Different scenarios for solutions to the Friedmann equation. If $\Lambda < 0$, the potential energy grows with a, and there is a maximum value for $a(t)$ for any given total energy (kinetic plus potential); the Universe expands to a maximum size and then starts to collapse again; a is always bounded. There is always a maximum value of $a(t)$ for $A > 0$ and $\Lambda < 0$. For $\Lambda > 0$ there is a maximum in the potential. If the Universe starts from a small size then, provided it has enough energy, it can get over the barrier; $a(t)$ will grow indefinitely. (This appears to correspond to our Universe.) If there is not enough energy to get over the barrier, the Universe expands for a while, reaches a maximum size, and then starts to contract again. If we start to the right of the maximum in the potential, at large a, with total energy below the maximum potential energy, then there is a minimum size below which we cannot go. If the energy is finely tuned to the correct value, the Universe can sit at the top of the potential barrier with no kinetic energy and it will stay there with constant a (labelled 'Einstein's static solution' in the figure), but this is clearly an unstable situation. (This was the solution originally posed by Einstein, before he knew about the Hubble expansion.) For *any* solution the energy is minus infinity for $a \rightarrow 0$.

Of course, $A > 0$. (There is some matter in the Universe!) And observationally K is negligible today, $\Omega_K \ll 1$, so we can get a good model of the current evolution of the Universe by putting K to zero. It is left as an exercise to show that the unique solution of (7.25) with $K = 0$ and initial condition $a(0) = 0$ is

$$a(t) = \left(\frac{3A}{\Lambda c^2} \right)^{1/3} \left\{ \sinh \left(\sqrt{\frac{3\Lambda}{4}} ct \right) \right\}^{2/3} . \tag{7.26}$$

Since $\sinh u \approx u$ for small u, the small t behaviour of this solution (with $ct \ll \frac{2}{\sqrt{3\Lambda}}$) is

$$a(t) = \left(\frac{9A}{4} \right)^{\frac{1}{3}} t^{\frac{2}{3}} \quad \Rightarrow \quad a^3(t) = \frac{9A}{4} t^2, \tag{7.27}$$

which is the exact solution of (7.25) for $K = \Lambda = 0$. Since $A = \frac{8\pi G}{3} \rho a^3$, this can be re-written as

$$\rho = \frac{1}{6\pi G t^2}. \tag{7.28}$$

This is called a *matter-dominated* Universe, and this was the behaviour of our Universe until quite recently in cosmological terms. The form (7.28) is notable because the speed of light does not appear; it is unchanged in the non-relativistic limit $c \to \infty$. The form $a^3 \propto t^3$ is somewhat reminiscent of Kepler's third law, and this is no coincidence. Although the physical situation is different (Kepler's third law refers to oscillatory behaviour and $a(t)$ in (7.27) is not periodic), there is an underlying physical reason for this similarity – the speed of light does not appear in Kepler's third law or in Equation (7.27) and, up to a dimensionless constant of proportionality, they can both be derived from dimensional analysis since G has dimensions of kg^{-1} m^3 s^{-2}.

At the opposite extreme, the large t behaviour of (7.26) leads to exponential growth:

$$a(t) \propto \exp \left(\frac{\sqrt{3\Lambda} ct}{2} \right).$$

Our Universe is in the process of transitioning from matter-dominated to exponentially expanding behaviour at the present time.

7.2 Microwave Background

We have seen that the Universe appears to have a finite age, and it happens that the Cosmic Microwave Background provides a unique window into the early Universe.

The Friedmann equation studied in the previous section did not include the effects of thermal radiation, because it is not relevant today, but it was more significant in the past. There is an energy density ϵ in thermal radiation which is proportional to the fourth power of the temperature T,

$$\epsilon = \frac{4\sigma}{c} T^4. \tag{7.29}$$

The value $\sigma = 5.67 \times 10^{-8}$ J s^{-1} m^{-2} K^{-4} is called the Stefan–Boltzmann constant. Thermal radiation contributes to the energy density of the Universe, and this should be included in T_{00}, which is then modified to

$$T_{00} = \rho c^2 + \epsilon \tag{7.30}$$

in Einstein's equations.

Focusing on thermal radiation alone for the moment, and ignoring matter, there is a non-zero pressure associated with radiation,

$$P = \frac{1}{3}\epsilon.$$

This is the equation of state for thermal radiation. Using this in (7.12), with ρ replaced by $\frac{\epsilon}{c^2}$, the conservation law reads

$$\frac{d}{dt}(\epsilon a^3) = -\frac{\epsilon}{3}\frac{da^3}{dt} \qquad \Rightarrow \qquad \frac{d(\epsilon a^4)}{dt} = 0,$$

hence $\epsilon \propto \frac{1}{a^4}$.

There is a physically intuitive way to understand this result. As $a(t)$ increases, comoving volumes increase like a^3. A characteristic wavelength can be associated with thermal photons by using the Einstein relation $E = h\nu$, via $h\nu = hc/\lambda = k_B T$, and their wavelength stretches like a, so their frequency decreases like $1/a$ as a increases (they are red-shifted). Their energy therefore decreases like $1/a$. Combining this with the volume increase, their energy density $\epsilon \propto \frac{1}{a^4}$. For future reference, note that this also implies that the temperature depends on time through

$$T \propto 1/a(t). \tag{7.31}$$

Denote the constant of proportionality as B in

$$\epsilon = \frac{3}{8\pi G}\frac{c^2 B}{a^4}.$$

Then, including the thermal radiation in T_{00} by using (7.30), the Friedmann equation (7.24) is modified to

$$\left(\frac{\dot{a}}{a}\right)^2 = \frac{A}{a^3} + \frac{B}{a^4} + \frac{\Lambda c^2}{3} - \frac{c^2 K}{a^2}. \tag{7.32}$$

The observed density of the Universe, from galactic dynamics, is $\rho = 2.5 \times 10^{-27} \mathrm{kg\,m^{-3}}$. For thermal radiation, $\epsilon = \frac{4\sigma T^4}{c}$, so $\frac{\epsilon}{c^2} = 4.642 \times 10^{-31} \mathrm{kg\,m^{-3}}$ with $T = 2.725$ K. So, at the present day,

$$\frac{\epsilon}{\rho c^2} = 1.9 \times 10^{-4}.$$

This means that today the energy density in CMB thermal photons is negligible compared to the density of matter, and so the thermal radiation was ignored in §7.1. But at earlier times,

$$\frac{\epsilon(t)}{\rho(t)c^2} = \frac{\epsilon(t_0)}{\rho(t_0)c^2}\frac{a_0}{a(t)} = (1.9 \times 10^{-4})\frac{a_0}{a(t)}.$$

For matter-dominated expansion (setting Λ and $K = 0$),

$$a(t) = \left(\frac{9A}{4}\right)^{1/3} t^{2/3} \propto t^{2/3},$$

where t is in seconds. Running the Friedmann equation backwards in time, the energy density in radiation equalled that in matter when

$$\left(\frac{t_0}{t}\right)^{2/3} = \frac{1}{1.9 \times 10^{-4}} \qquad \Rightarrow \qquad t = (1.9\times 10^{-4})^{3/2}t_0 = 2.6 \times 10^{-6}\, t_0,$$

a few millionths of its present age. With the best fit for the age of the Universe $t_0 = 13.8 \times 10^9 \mathrm{yr}$, we get that $\frac{\epsilon}{c^2} = \rho$ when $t = 36{,}000\,\mathrm{yr}$. For $t < 36{,}000\,\mathrm{yr}$ the energy density in radiation dominated over that in matter, so for early times it is safe to ignore A, and the Friedmann equation becomes

$$\left(\frac{\dot{a}}{a}\right)^2 = \frac{B}{a^4} + \frac{\Lambda c^2}{3}.$$

But we also have $\frac{B}{a^4} \gg \frac{\Lambda c^2}{3}$ when $t < 36{,}000\,\mathrm{yr}$ (though not today), so in the early universe ($t < 10^4 \mathrm{yr}$) a very good approximation is

$$\dot{a}^2 = \frac{B}{a^2}$$

$\Rightarrow \dot{a}a = \sqrt{B} \Rightarrow \frac{1}{2}\frac{d(a^2)}{dt} = \sqrt{B} \Rightarrow a^2 = 2\sqrt{B}t + const.$ With initial condition $a(0) = 0$ we get

$$a(t) = (2\sqrt{B})^{1/2}t^{1/2}.$$

In summary, we can divide the history of the Universe into three principal epochs:

Summary

$$t > t_0 = 10^{10}\text{yr} \qquad \left(\frac{\dot{a}}{a}\right)^2 = \frac{\Lambda c^2}{3} \qquad a(t) \propto \exp\left(\sqrt{\frac{\Lambda}{3}}ct\right) \qquad \begin{array}{l}\text{exponential} \\ \text{expansion}\end{array}$$

$$36{,}000\,\text{yr} < t < t_0 \qquad \left(\frac{\dot{a}}{a}\right)^2 = \frac{A}{a^3} \qquad a(t) \propto t^{2/3} \qquad \text{matter dominated}$$

$$t < 36{,}000\,\text{yr} \qquad \left(\frac{\dot{a}}{a}\right)^2 = \frac{B}{a^4} \qquad a(t) \propto t^{1/2} \qquad \text{radiation dominated;}$$

$\frac{K}{a^2}$ never was and never will be significant. If we were geologists, we would have arcane names for these three epochs; but physicists just talk about radiation-dominated, matter-dominated, and Λ-dominated.

7.3 Thermal History of the Early Universe

Since the Universe is expanding, the cosmic scale factor $a(t)$ was smaller at earlier times than it is today and, from Equation (7.31), it was hotter when it was younger. Currently the size of the visible Universe is approximately $a_0 \approx 10^{26}$m and the temperature is $T \approx 3\,\text{K}$. At $t = 36{,}000\,\text{yr}$, $\frac{a_0}{a} = 10^4$ so the temperature was $T \approx 3 \times 10^4\,\text{K}$. At these early times, there were no stars or galaxies; all the baryonic matter in the Universe (mostly hydrogen and helium at that time) was in gaseous form, and at these temperatures all that gas was ionised. The ionisation energy of hydrogen is 13.6 eV and 1 eV is $1.6 \times 10^{-19}\,\text{J}$, so 1 eV is equivalent to a temperature of 11,600K, given by $k_B T = 1.6 \times 10^{-19}\,\text{J}$. We might therefore expect that all the hydrogen gas in the Universe will ionise at a temperature of about $13.6 \times 11{,}600 = 160{,}000\text{K}$. In the early Universe all the hydrogen is actually ionised at a much lower temperature, because the number density of photons is much higher than the number density of protons; there are enough photons in the upper end of the tail of the black-body distribution (see Figure 7.5) to ionise all the hydrogen in the Universe; the peak of the spectrum can be a lot less than 160,000 K and all the protons can still be ionised. At the present day, the mass density of baryonic matter is $\approx 4.5 \times 10^{-28}\,\text{kg m}^{-3}$; dividing by the mass of the proton, 1.67×10^{-27} kg (almost the same as the mass of a neutron) gives an estimate for the present-day number density of baryons,

$$n_B \approx \frac{4.5 \times 10^{-28}}{1.67 \times 10^{-27}} \approx 0.3 \, \mathrm{m}^{-3};$$

on average, there is about one baryon in every three cubic metres of space. An estimate of the present-day number density of photons, at a temperature of 2.7 K, is given by dividing the energy density in the present-day microwave background,

$$\epsilon = \frac{4\sigma T^4}{c} = 4 \times 10^{-14} \mathrm{J \, m}^{-3},$$

by the average energy of the thermal photons, $k_B T$ with $T = 2.7$ K,

$$\frac{\epsilon}{k_B T} \approx 10^9 \, \mathrm{m}^{-3}.$$

(Roughly speaking, a typical photon in a bath of thermal radiation at $T = 2.7$ K has a wavelength of about $\lambda = 1$ mm, and there is one photon in a volume of $\lambda^3 = 10^{-9} \mathrm{m}^3$.) Using these gives the dimensionless baryon/photon ratio:

$$\eta = \frac{n_B}{n_\gamma} \approx 3 \times 10^{-10}. \tag{7.33}$$

At the present day, there are more than a billion photons for every baryon and, since both n_B and n_γ are proportional to $1/a$, we can assume that the baryon-to-photon ratio, $n_B/n_\gamma \approx 10^{-9}$, has been constant at least back to $t = 36,000$ yr and even further.

A careful calculation shows that all the hydrogen in the Universe was ionised when the temperature of the Universe was about 4,000 K. The temperature reached 4,000K when

$$\frac{a_0}{a} \approx 10^3 \Rightarrow \frac{t_0}{t} = 10^{9/2} = 3 \times 10^4 \Rightarrow t = \frac{1.4 \times 10^{10} \mathrm{yr}}{3 \times 10^4} = 500,000 \, \mathrm{yr}.$$

(A more accurate figure is $t_s = 380,000$ yrs.) Before this time, all the baryonic matter in the Universe was an ionised plasma consisting of ionised hydrogen and helium with free electrons; afterwards it is mostly neutral hydrogen and helium until some of it gets re-ionised due to the heat from stars that form a few hundred million years later. A plasma is opaque, and the mean free path of a photon was very short when the Universe was less than 380,000 years old – there was a thick fog.

The fog suddenly cleared at 380,000 years, and thereafter most of space was clear until galaxies started forming at about a few hundred million years, and even then most of intergalactic space was still transparent. At 380,000 years, the plasma of protons and electrons condensed, mostly into neutral hydrogen, and light could travel freely thereafter. The microwave background that we see in Figure 7.2 consists of photons that have travelled towards us in straight lines, unimpaired, since 380,000 years after the Big Bang. This is called the *surface of last scattering*.

However, the temperature of the microwave background is not completely uniform, although it is extremely smooth (see Figure 7.2). The average is 2.725 K, but there are fluctuations; the temperature depends slightly on the direction, with variations $\Delta T(\theta, \phi)/T$ of the order of a few times 10^{-5}. We can expand $\Delta T(\theta, \phi)$ in spherical harmonics, $Y_{lm}(\theta, \phi)$,

$$\Delta T(\theta, \phi) = \sum_{l=0}^{\infty} \sum_{m=-l}^{m=l} (\Delta T_{lm}) Y_{lm}(\theta, \phi),$$

and extract the harmonic coefficients ΔT_{lm} using orthogonality of the $Y_l^m(\theta, \phi)$,

$$\Delta T_{lm} = \int_0^{\pi} \int_0^{2\pi} \Delta T(\theta, \phi) \{Y_{lm}(\theta, \phi)\}^* \sin\theta d\theta d\phi.$$

The data are most easily visualised by summing over m and defining

$$C_l = \frac{1}{2l+1} \sum_{m=-l}^{l} |\Delta T_{lm}|^2.$$

These give an idea of the magnitude of the temperature fluctuations at an angular size $\theta \approx \frac{\pi}{l}$ and are shown in Figure 7.5, which is called the CMB power spectrum.

The peaks are due to the fact that, in the plasma phase before 380,000 years, electrons, protons, and α-particles (doubly ionised helium) were strongly coupled to photons, and plasma oscillations were imprinted on the microwave background; the angular size (l) is a measure of the wavelength of the plasma oscillation – the wavelength of sound, cosmic sound. Since protons are baryons, these peaks are called *Baryon Acoustic Oscillations* (BAOs). We can learn a lot about the physical conditions in the Universe at $t \approx 380,000$ years from the shape of these peaks. The separation of the peaks gives information about the value of K ($K = 0$ is the

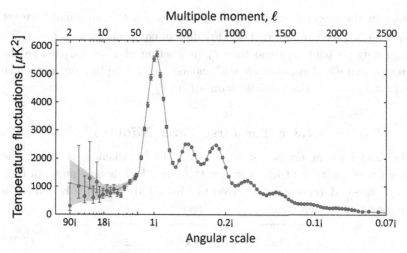

Figure 7.5 **The Cosmic Microwave power spectrum.** The power spectrum (actually $l(l+1)C_l$ for clarity) as a function of angular size l, in units of μK^2. The dots are the data points; the solid line is the best fit using the cosmological model described here, with Ω_M and Ω_Λ given on page (152). ©ESA/Planck Collaboration

best fit), and the height of the peaks gives information about the density of protons and α-particles, the baryonic density, which is compatible with the previously mentioned present-day figure of $\approx 4 \times 10^{-28}$ kg m^{-3}. With some extra input, H_0 can be estimated from these fluctuations in the CMB temperature, and the best fit is

$$H_0 = 67.27 \pm 0.60 \, \text{km s}^{-1} \text{Mpc}^{-1} \quad \Rightarrow \quad h = 0.6727 \pm 0.0060. \quad (7.34)$$

This is in some tension with the value 73.2 in (2.30) obtained from redshift luminosity observations (Figure 2.6), and there is currently much debate about this discrepancy. Are the errors underestimated or is this an indication that something is missing from the Friedmann equation (7.32)? The value (2.30) is obtained from observations around $z \approx 1$, while (7.34) uses data from $z \approx 1,000$, so we are relying on the Friedmann equation over three orders of magnitude in z. Perhaps the numbers will be reconciled as the data improve or perhaps this is a signal of something missing from our understanding; time will tell.

We can follow the history of the early Universe back to times well before 380,000 years, and there is more fascinating physics to be discovered here, which will be explored in more detail in the next section;

but for the moment we just observe that as $a \propto t^{1/2}$ for small t we get $\dot{a} \xrightarrow{t \to 0} \infty$. We cannot trust the Friedmann equation back to $t = 0$. We can only go back to some time t_μ (a fraction of a second) where the temperature and energies are still understood and replace our ignorance of $t < t_\mu$ with initial conditions on $a(t_\mu)$.

7.3.1 The First Three Minutes

Looking back at times $t < 36,000 \, \text{yr} \approx 10^{12} \, \text{s}$, when $T = 10^4 \, \text{K}$, the scale factor $a(t)$ went like $a \propto \frac{1}{T} \propto t^{1/2}$. Now $k_B T$ is an energy, in fact, $k_B T \approx \frac{1}{\sqrt{t}}$ if the energy is given in MeV ($1 \, \text{MeV} \approx 10^{10} \, \text{K}$) and t in seconds, so we can write

$$\boxed{T \, (10^{10} \, \text{K}) \approx T \, (\text{MeV}) \approx \frac{1}{\sqrt{t \, (\text{secs})}}} \qquad (7.35)$$

It is remarkable how much we can deduce about the early Universe from this equation.[6]

At the age of a few times 10^4 years, the Universe contained an over-all electrically neutral plasma of ionised hydrogen (protons, p), helium (mostly ^{4}He but some ^{3}He) with a hint of deuterium (^{2}H, the isotope of hydrogen with one proton and one neutron in the nucleus), and lithium, together with electrons, photons, and neutrinos, all in thermal equilibrium. There was also non-baryonic dark matter, but we know so little about this that we cannot say much about its physical properties. How-ever, for $t < 36,000$ years, dark matter was not relevant for the dynamical evolution of the cosmological scale factor $a(t)$, and we shall simply leave it out of the discussion. No doubt, interesting things were happening in the dark sector at early times, but we cannot say anything about it with the current status of our knowledge.

The binding energy of neutrons and protons in a helium nucleus is about $100,000 \, \text{eV}$, or $0.1 \, \text{MeV}$, corresponding to 10^9 K, and at these temperatures helium nuclei disintegrate into protons and neu-trons. According to Equation (7.35) this corresponds to $100 \, \text{s}$ after the Big Bang and earlier. Before this time there was no helium or any other heavier elements in the Universe, only protons and neu-trons, together with electrons, photons, and neutrinos. After this time,

---

[6] The title of this section is taken from the popular book by Steven Weinberg, which is a very readable account of some of the physics described here.

there were essentially no free neutrons; they were all bound into heavy hydrogen (deuterium, ^{2}H) and helium nuclei (with a little bit of lithium) – this is known as the *era of nucleosynthesis*.[7]

In free space, when not bound into an atomic nucleus, neutrons decay to protons via β-decay,

$$n \to p + e^- + \bar{\nu},$$

with a lifetime of about $\tau = 900$ s. Neutrons are about $1.3\,\mathrm{Mev}/c^2$ heavier than protons, and inverse β-decay,

$$p + e^- \to n + \nu,$$

requires a minimum amount of energy to proceed, but it can occur down to temperatures slightly below $10^{10}\,\mathrm{K}$, corresponding to a time 2 s after the Big Bang. Before 2 s, neutrons and protons were in thermal equilibrium, and this process could go both ways:

$$p + e^- \leftrightarrow n + \nu.$$

After 2 s, neutrons and protons are no longer in thermal equilibrium, and between 2 s and 100 s, free neutrons can decay to protons, $n \to p + e^- + \bar{\nu}_e$. After 100 s, neutrons are mostly bound into helium nuclei and are stable, because the only energy levels available to a proton arising from neutron decay in the nucleus are blocked, due to the Pauli exclusion principle, by the other two protons already present in the helium nucleus. We can estimate how many neutrons are left at 100 s, using the lifetime for neutron decay in free space. We should allow for the fact that the neutron-to-proton ratio N_n/N_p is not quite unity at 2 s, due to the neutron-proton mass difference $\Delta E := (m_n - m_p)c^2 = 1.3$ MeV. The Maxwell–Boltzmann distribution for particles of energy E in a gas at temperature T is $n(E) \sim e^{-E/k_B T}$. Using $T = 1/\sqrt{t}$, with T measured in MeV and t in seconds gives, in thermal equilibrium,

$$\frac{N_n}{N_p} = e^{-\Delta E \sqrt{t}} \quad \Rightarrow \quad N_n = N_p e^{-\Delta E \sqrt{t}}.$$

If neutrons drop out of thermal equilibrium at a time t_e and subsequently decay to protons until they are bound into helium nuclei at time t_{He}, then we expect the neutron-proton ratio at t_{He} to be

---

[7] Heavier elements, such as carbon, nitrogen, and oxygen, were formed in stars much later, and even heavier elements, such as gold, require supernovae.

$$\frac{N_n}{N_p} = e^{-\frac{(t_{He}-t_e)}{\tau}} e^{-\Delta E \sqrt{t_e}}.$$

Using the values $\Delta E = 1.3$ MeV, $\tau = 900$ s, $t_e = 2$ s and $t_{He} = 100$ s gives

$$\frac{N_n}{N_p} \approx \frac{1}{7},$$

which is consistent with the observed ratio of hydrogen to helium in very old stars.

If the rate of helium production in the early Universe were larger, more helium would be produced at earlier times, absorbing neutrons, and the neutrons would have less time to decay, resulting in a larger N_n/N_p ratio and hence a larger He/H ratio. If the rate of helium production were lower, neutrons would remain free for longer and have more time to decay, resulting in a smaller N_n/N_p ratio and hence smaller He/H ratio. The rate for helium production increases if the density of neutrons and protons increases, and the observed ratio of primordial hydrogen to helium in the Universe puts a limit on the maximum allowed density of neutrons and protons consistent with observations. At the present day, baryonic matter cannot be more than 15 per cent of the total amount of matter in the Universe, otherwise the rate of helium production in the early Universe goes up and the observed ratio of the amount of hydrogen relative to the amount of primordial helium would be higher than observed. There are similar considerations for other light elements that were produced in the Big Bang – deuterium, ^{3}He, and ^{7}Li; see Figure 7.6.

Before nucleosynthesis started, at 4 s after the Big Bang, the temperature was 5×10^9 K, the energy equivalent of 0.5 Mev, and this is significant because the rest mass of an electron is 0.511 Mev/c^2. This means that two thermal photons at 5×10^9 K that collide have enough energy to produce an electron–positron pair

$$2\gamma \rightarrow e^- + e^+.$$

At these temperatures and above, the Universe is swamped by electron–positron pairs, in much greater abundance than the number of electrons that we see today.

We can trace the thermal history of the Universe even further back: at temperatures equivalent to about 150 MeV ($\sim 10^{12}$ K) protons and neutrons evaporate into quarks and gluons,and it is believed that at a

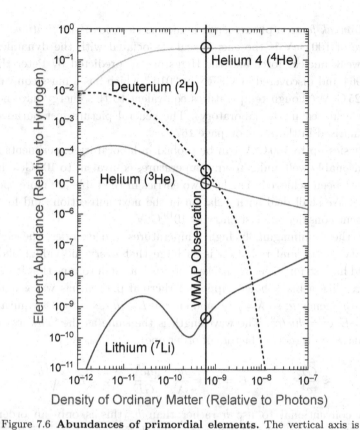

Figure 7.6 **Abundances of primordial elements.** The vertical axis is the abundance of each isotope relative to hydrogen. The top curve shows the mass fraction of primordial ^{4}He relative to hydrogen; the measured ^{4}He abundance lies between 0.23 and 0.25. Notice that the theoretical prediction for ^{4}He increases as the baryon density (the density of protons and neutrons) increases, as described in the text. The tightest constraints come from observations of inhomogeneities in the cosmic microwave background and Baryon Acoustic Oscillations from the Wilkinson Microwave Anisotropy Probe (WMAP – the vertical band) indicating a value for the baryon density of $(4.1 \pm 0.1) \times 10^{-28}$ kg m^{-3}, putting the density of protons plus neutrons at about 15 per cent of the total mass density, implying the existence of another, unknown, type of matter.
Figure from NASA/WMAP Science Team.

temperature equivalent to an energy of about $100\,\mathrm{GeV}$ ($\sim 10^{15}\,\mathrm{K}$) the electromagnetic and the weak nuclear forces are no longer really distinct forces but should be thought of as different aspects of a single force, the *electroweak force*, a phenomenon known as *electroweak unification*.[8] The figure of 100 GeV is the energy scale associated with the dynamics of the weak nuclear force and the Higgs boson, predicted by Peter Higgs in 1964 and discovered at CERN in 2016,[9] which has a mass equivalent to 125 GeV (though temperatures equivalent to this energy have never been achieved in the laboratory). The general picture just outlined is summarised in the table on page 167.

Physics up to 100 GeV can be probed in laboratory experiments and is reasonably well understood; temperatures equivalent to 100 GeV have not yet been achieved. The last two entries in the table are more speculative. We shall deal with inflation in the next subsection, and for the moment consider the last entry at 10^{19} GeV.

At these unimaginably high temperatures and energies, the energy density in thermal radiation is so large that every individual photon would have so much energy packed into such a small volume that it would behave like a black hole. Suppose a thermal photon has wavelength λ with frequency ν, so $\lambda = c/\nu$, and energy $E = h\nu$, with mass equivalent $M = E/c^2 = h\nu/c^2$. The wavelength is the same as the Schwarzschild radius when (ignoring factors of order one)

$$\lambda \approx \frac{GE}{c^4} \quad \Rightarrow \quad E = h\nu = \frac{hc}{\lambda} = \frac{hc^5}{GE} \quad \Rightarrow \quad E = \sqrt{\frac{hc^5}{G}}.$$

It is conventional to use $\hbar$ rather than h (this is only an order of magnitude estimate) and define the *Planck Energy* as

$$\boxed{E_{Planck} = \sqrt{\frac{\hbar c^5}{G}} = 2 \times 10^9 \text{ J} = 1.2 \times 10^{19} \text{ GeV}}$$

and the mass equivalent is the *Planck Mass*,

$$M_{Planck} = E_{Planck}/c^2 = \sqrt{\frac{\hbar c}{G}}$$

which is about $10^{19} \, m_{proton}$, 10^{19} Gev/c^2, or 10^{-5} gm. This is an incredibly high energy for a single fundamental particle.

---

[8] Unlike the unification of electricity and magnetism achieved in Maxwell's equations, which produced the relation $\epsilon_0 \mu_0 = \frac{1}{c^2}$, the number of free physical parameters is not reduced in electroweak unification; rather, electromagnetism and the weak nuclear force are intimately mixed together at energies of 100 GeV and higher.

[9] Peter Higgs received the Nobel Prize in 2013 for his prediction, sharing it with François Englert.

Time	Temperature/ Energy	
Exponential ↑ expansion		$a \propto e^{\sqrt{\frac{\Lambda}{3}}ct}$
$1.38y \times 10^{10}\,\text{yr}$	2.7K	t_0, present day
Matter ↓ dominated		$a \propto t^{2/3}$
$4.56 \times 10^9\,\text{yr}$	6K	Solar System formed
$10^9\,\text{yr}$	15K	Era of active galaxy formation: quasars and active galactic nuclei, heavy elements created in stars and supernovae
$10^8\,\text{yr}$	70K	First stars and galaxies formed
$380,000\,\text{yr}$	4,000K	Hydrogen ionises, surface of last scattering
Matter ↑ dominated		$a \propto t^{2/3}$
Radiation ↓ dominated		$a \propto t^{1/2}$
$3.6 \times 10^4\,\text{yr}$	10,000K	No elements higher than Li in the periodic table; plasma of p, $^2\text{H}^+$, $^4\text{He}^{2+}$, $^3\text{He}^{2+}$, $^7\text{Li}^{3+}$, e^-, γ, ν, $\bar\nu$, 75% p and 25% ^4He by mass
100s	(10^9K) 0.1 MeV	Era of nucleosynthesis ^4He forms as do ^3He, ^2H, and Li, $N_n/N_p = 1/7$
4s	0.5 MeV	γ's in thermal background can produce e^+, e^- pairs
1s	$(5 \times 10^9\text{K})$ 1 MeV	$p, n, e^+, e^-, \nu, \bar\nu, \gamma$ β-decay, $n \to p + e^- + \bar\nu$, starts to deplete neutrons
$10^{-2}\,\text{s}$	10 MeV (10^{11}K)	$p + e^- \leftrightarrow n + \nu$ bi-directional protons and neutrons in thermal equilibrium, $N_p/N_n = 1$
$5 \times 10^{-5}\,\text{s}$	150 MeV $(1.5 \times 10^{12}\text{K})$	Free quarks and gluons condense into protons and neutrons Plasma of free quarks, anti-quarks, $e^-, e^+, \mu^-, \mu^+, \nu, \bar\nu, \gamma$ and gluons
$4 \times 10^{-6}\,\text{s}$	500 MeV $(5 \times 10^{12}\text{K})$	Highest temperatures reached in experiments to date
$10^{-10}\,\text{s}$	100 GeV	Electromagnetism and weak nuclear force unify into the electroweak force
$10^{-14}\,\text{s}$	10,000 GeV	Highest energies so far achieved in experiments at CERN
$10^{-36} \sim 10^{-32}\,\text{s}$	$10^{15} \sim 10^{13}$ GeV	Era of inflation
$10^{-43}\,\text{s}$	10^{19} GeV	Era of Quantum Gravity; Friedmann equation must fail

Dividing the Planck energy by Planck's constant gives one over the *Planck Time*,

$$t_{Planck} = \sqrt{\frac{\hbar G}{c^5}} \approx 10^{-43} \text{ s},$$

and multiplying this by the speed of light gives the *Planck Length*,

$$l_{Planck} = \sqrt{\frac{\hbar G}{c^3}} \approx 10^{-35} \text{ m}.$$

The Friedmann equation cannot be trusted when the energy density approaches the Planck energy divided by the Planck volume (the Planck length cubed). At these fantastic energy densities, Einstein's general theory of relativity cannot be the correct framework for describing physics, and quantum effects probably require some new, as yet unknown, quantum theory of gravity. This does not mean that general relativity is wrong, any more than general relativity implies that Newtonian gravity is wrong. Newtonian gravity is correct and is a hugely successful theory, but it has a limited range of applicability, and it has to be modified when speeds approach the speed of light. Similarly, general relativity is an immensely successful theory, which reduces to Newtonian gravity in the non-relativistic limit $c \to \infty$, with a huge amount of observational support, but it must also have a limited range of applicability and almost certainly needs to be modified at time scales near and below the Planck time and energy densities near and above the Planck energy density.

In fact, the Planck energy appears in the Friedmann equation naturally, in much more modest conditions. Using the explicit expression for the Stefan–Boltzmann constant,

$$\sigma = \frac{\pi^2 k_B^4}{60 \hbar^3 c^2},$$

the energy density in thermal photons is

$$\epsilon_{Rad} = \frac{4}{c} \sigma T^4 = \frac{\pi^2}{15 c^3 \hbar^3} (k_B T)^4,$$

equivalent to a mass density of

$$\frac{\epsilon}{c^2} = \frac{\pi^2}{15 c^5 \hbar^3} (k_B T)^4.$$

At early times, when the Universe was less than about 36,000 years old, $\frac{\epsilon}{c^2}$ dominates the Friedmann equation,

$$H^2 = \frac{8\pi G}{3} \frac{\epsilon}{c^2} = \frac{8\pi^3}{45} \frac{G}{c^5 \hbar^3} (k_B T)^4 = \frac{8\pi^3}{45 \hbar^2} \frac{(k_B T)^4}{(E_{Planck})^2}.$$

Then the cosmic scale factor $a(t) \propto t^{1/2}$, so $H = \frac{1}{2t}$, and[10]

$$(k_B T)^4 = \frac{45}{32\pi^3} \left(\frac{\hbar E_{Planck}}{t} \right)^2.$$

This formula is valid for times as late as 36,000 years. But the appearance of the Planck energy here is not necessarily a signal of quantum gravity effects; it is merely due to the fact that a classical gravitational field is being sourced by quantum matter (thermal photons).

However, the Friedmann equation cannot be trusted for times close to and earlier than 10^{-43} s after the Big Bang; but there are other unknowns long before we hit this quantum barrier of ignorance. The highest temperature ever achieved in a laboratory to date is 5×10^{12} K (equivalent to an energy close to 500 MeV), produced in heavy ion collisions in ALICE (A Large Ion Collider Experiment) at CERN in Geneva, a temperature corresponding to about 4 μs after the Big Bang; and the highest energy so far achieved in a laboratory, without thermal equilibrium, is 13 TeV (1.3×10^7 MeV), again at CERN. We have never explored energies higher than this and so cannot be sure what lies up there, though we already know there must be non-baryonic dark matter and there could be other unknown forms of matter, or unknown physics, affecting the time evolution of the Friedmann equation.

But there is another problem with the Friedmann equation in the early Universe that stems from a different source – the horizon problem.

7.3.2 The Horizon Problem and Inflation

Microwave photons from two points 180° apart in the sky come from the 'surface of last scattering' when neutral hydrogen was formed at $t_s \approx 380{,}000$ yr at $T \approx 4{,}000$ K. Two diametrically opposite points in the sky appear to be the same temperature (to within a few parts in 10^5), so presumably they were in thermal contact at some point in the past. But this is inconsistent with our model.

To understand that there is a problem, first choose coordinates so that our own Galaxy sits at $r = 0$ and we observe photons coming in from the surface of last scattering at a coordinate distance r_s. In a short time interval dt at an intermediate time t, a photon travels a physical

---

[10] Actually, as written, this equation is only valid for temperatures for which $k_B T \ll m_e c^2$, that is, times later than about 4 s. We have already seen that, for temperatures of order $2m_e c^2/k_B$ and greater, electron–positron pairs can be created out of thermal energy and they contribute to $\frac{\epsilon}{c^2}$. This modifies the prefactor $\frac{8\pi^3}{45\hbar^2}$, but the general conclusion is unchanged.

distance $-a(t)dr = cdt \Rightarrow dr = -c\frac{dt}{a(t)}$ ($dr < 0$ because the photon is travelling inwards, towards us):

$$\int_{r_s}^0 dr = -c \int_{t_s}^{t_0} \frac{dt}{a(t)}.$$

The 'metre stick' radius of the surface of last scattering at t_s is[11]

$$d_s = a(t_s)r_s = ca(t_s) \int_{t_s}^{t_0} \frac{dt}{a(t)}.$$

For $t_s < t < t_0$ we use $a(t) = bt^{2/3}$, with b a constant. Therefore,

$$d_s = a(t_s)\frac{c}{b} \int_{t_s}^{t_0} \frac{dt}{t^{2/3}} = bt_s^{2/3}\frac{c}{b} \left[3t^{1/3}\right]_{t_s}^{t_0}$$

$$= 3ct_s^{2/3} \left(t_0^{1/3} - t_s^{1/3}\right) = 3ct_s \left[\left(\frac{t_0}{t_s}\right)^{1/3} - 1\right].$$

With $t_0 \approx 1.4 \times 10^{10}$yr and $t_s \approx 3.8 \times 10^5$yr, $\left(\frac{t_0}{t_s}\right) \approx 3{,}700$ and

$$d_s \approx 3c \times (3.8 \times 10^5\text{yr}) \times 32 \approx 3 \times 10^{23} \text{ m}.$$

Let the distance a photon could have travelled at time t_s, since $t = 0$, be d_h. For $t < t_s$ use $a(t) = b't^n$ ($n = \frac{2}{3}$ for matter-dominated and $n = \frac{1}{2}$ for radiation-dominated). Now

$$d_h = cb't_s^n \int_0^{t_s} \frac{dt}{b't^n} = ct_s^n \left[\frac{t^{-n+1}}{-n+1}\right]_0^{t_s} = \frac{c}{1-n}t_s,$$

for example, for $n = \frac{1}{2} \Rightarrow d_h = 2ct_s$, but then

$$\frac{d_s}{d_h} = \frac{3}{2} \left[\left(\frac{t_0}{t_s}\right)^{1/3} - 1\right] \approx 50.$$

How can two points on the surface of last scattering, $180°$ apart in the sky and therefore a distance $2d_s$ apart, be in thermal equilibrium with each other, when $d_s \approx 50d_h$ is 50 times the distance a photon could have travelled since the beginning of the Universe? This is known as 'the Horizon Problem' (see Figure 7.7). The situation is even worse if we demand thermal equilibrium for times $t < t_s$, as we do for nucleosynthesis at $t = 100$ s, since then $\left(\frac{t_0}{t}\right)^{1/3} > 33$. For example, $\left(\frac{t_0}{100\text{s}}\right)^{1/3} \approx 160{,}000$.

The most extreme case of the horizon problem is to assume thermal equilibrium at the Planck time, $t_{Planck} \approx 10^{-43}$ s. The Planck energy, $E_{Planck} \approx 10^{19}$ GeV, is equivalent to a temperature $T_{Planck} \approx 10^{32}$ K.

---

[11] The metre stick distance was defined on page 150.

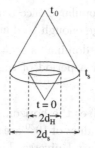

Figure 7.7 **Surface of last scattering.** The surface of last scattering is emitting photons towards us from a sphere of radius d_s, but two points on that sphere can only be in thermal equilibrium if they are within a distance d_h of each other, otherwise there has not been enough time since the Big Bang at $t = 0$ for energy to be shared between the two points.

If we replace our ignorance about $t = 0$ by the assumption that the Universe started off at the Planck time, with the Planck temperature, and assume that the subsequent expansion was radiation dominated, $a(t) \propto \sqrt{t}$ according to the Friedmann equation, and that the temperature drops as $T \propto 1/a(t) \propto 1/\sqrt{t}$, then at the present day, $t_0 = 4 \times 10^{17}$s, the temperature would be 50 K. Allowing for the changeover from $t^{1/2}$ to $t^{2/3}$ behaviour of $a(t)$ at 36,000 yr brings this down to ≈ 5 K, remarkably close to the observed present-day temperature of 2.7 K. However, if we also assume that the Universe started off at the Planck size, $l_{Planck} \approx 10^{-35}$ m. then its present size would only be about a millimetre (the wavelength of a present-day microwave photon). But the size of the observable Universe is $\approx 10^{26}$ m, too large by a factor of 10^{29}.

Inflationary Universe
Possible solutions:

1. The Friedmann equation is wrong – it breaks down for some time $t \ll 10^{-10}$ s (this must happen at some early time anyway, since we cannot allow $\dot{a} \to \infty$), for example, quantum gravity.
2. The Friedmann equation is correct, but change the matter content, introduce some new form of matter at very early times that drops out of the equation sometime before 10^{-15} s, so as not to disturb nucleosynthesis and the physics that we understand at 10 TeV and lower. In the 'inflationary universe' picture, it is assumed that a very large positive cosmological constant 'switched on' for a very brief period

at a very early time (a popular choice among workers in the field is $t \sim 10^{-35}$ s) and was large enough to dominate the dynamics of the Universe's evolution. If Λ dominates $\dot{a}^2 = \frac{\Lambda c^2 a^2}{3}$ and $a = a(0)e^{\sqrt{\frac{\Lambda}{3}}ct}$ exponential expansion for a period of time, between t_1 and t_2, say, with $t_1 < t_2$.

The second hypothesis solves the horizon problem if the inflationary factor $a(t_2)/a(t_1) \gg 1$, and values of 10^{30} are not uncommon in mathematical models of the inflationary Universe. The reason for this very early period of exponential expansion is usually attributed to a new kind of field, with its associated particles, called the 'inflaton', for which there is no observational evidence; its existence is postulated purely from a desire to solve the horizon problem. If the inflaton exists, it is possible that it is related to non-baryonic dark matter, but at the moment this is only speculation.

As a by-product of inflation, we get a natural explanation of why $K \approx 0$. Non-zero K is associated with curvature of 3-dimensional space. For example, positive curvature, $K > 0$, results in parallel lines converging, like great circles intersecting on the surface of a 2-dimensional sphere. The greater the radius of the sphere, the smaller the curvature; it is difficult to detect the curvature of the Earth's surface on lengthscales of a few metres. If K is non-zero before inflation, its significance after inflation is greatly reduced by the expansion of 3-dimensional space during the inflationary period, and the relevance of K in the Friedmann equation becomes negligible for all times after the period of inflation has finished. This could explain why attempts to measure K today give a null result.

7.3.3 Summary

The scenario just described is an immensely satisfying self-consistent view of how everything started 13.8 billion years ago and the elements came into being from the unimaginably hot plasma that was the very early Universe. The Friedmann equation cannot be trusted all the way back to $t = 0$, but it is in good shape at least as far back as 10^{-10} seconds after the Big Bang and there is no obvious reason why it cannot be trusted to include new physics such as an inflationary period as far back as 10^{-36} s; but what started it all is still a mystery. One unanswered question, crucial to our existence, is why the baryon-to-photon ratio (7.33), 10^{-9}, is so small? Where does this number come from? We are made of matter, not anti-matter; there is more matter than anti-matter in the Universe, but there is no obvious reason why this should be so. A microsecond after the Big Bang, the Universe was a seething plasma of particles and anti-particles, quarks and anti-quarks, electrons and positrons,[12] as well as non-baryonic dark matter, in whatever form that takes. By one second after the Big Bang, almost all of the anti-quarks and positrons had annihilated with quarks and electrons into photons, leaving a relatively small number of quarks and electrons to make all of the galaxies, stars, and planets that we see today. It is not currently known whether non-baryonic dark matter is a mixture of matter and anti-matter, or if it is predominantly matter, but the density of non-baryonic dark matter is within a factor of 10 of the baryonic matter density; is there some reason why these values are so close?

In 1967, the Soviet physicist Andrei Sakharov proposed a number of criteria necessary for a non-zero baryon density to emerge from an early Universe that starts off with no imbalance between the number of particles and anti-particles. One of these is that there must be an asymmetry between the behaviour of particles and anti-particles in the underlying dynamical interactions of fundamental particles. (Otherwise there would be the same number of anti-protons as protons in the present-day Universe, and this is not the case.) Such an asymmetry has been observed in particle decays in the laboratory, but the observed magnitude is not enough to explain the observed baryon-to-photon ratio; the standard model of particle physics predicts that the baryon-to-photon ratio in Equation (7.33) should be significantly *less* than 10^{-9}. The source of the observed preponderance of matter over anti-matter in our Universe is not understood and requires new physics beyond that of the standard model of particle physics. It is almost certainly related to the nature of non-baryonic dark matter.

The very early Universe is a fascinating place where cosmology and particle physics meet and intimately intertwine to produce a truly beautiful view of our world; but the picture is not complete and raises many questions:

- What is non-baryonic dark matter? Is it just a single type of particle that must simply be added to the pantheon of particles in the

---

[12] And some other fundamental particles that have been seen in the laboratory and are incorporated into the standard model of particle physics but have not been mentioned here. These all eventually decay to leave just up and down quarks that form protons and neutrons, electrons, neutrinos, and photons.

standard model of particle physics, the quarks and leptons that we already know of? Or is there a whole new zoo of unknown particles out there, interacting with each other in complicated ways and forming emergent macroscopic objects, like dark stars and dark galaxies, that are simply invisible to us except through their gravitational effects?

• What is the source of the preponderance of baryonic matter over anti-matter?

• What happens at the Planck temperature when energy densities approach the Planck value, or at curvature singularities like the singularity at $r = 0$ in the Schwarzschild metric? Einstein's general theory of relativity must break down under these extreme conditions and we need some completely new idea here.

Furthermore, the current epoch is a curious time when the Universe is just switching over from matter dominated to exponential expansion, and the future looks like an ever-expanding Universe and eventual heat death, as described in page 58. Is that what our future holds?

Problems

1) Evaluate the connection coefficients and the components of the Riemann tensor for the Robertson–Walker line element in Appendix E,

$$ds^2 = -c^2dt^2 + a^2(t)\left(dx_1^2 + dx_2^2 + dx_3^2\right) = g_{\mu\nu}dx^\mu dx^\nu.$$

Determine the Einstein tensor and the differential equations that $a(t)$ must obey for Einstein's equation to be satisfied for a homogeneous fluid at rest with energy-momentum tensor

$$T_{\mu\nu} = \begin{pmatrix} \rho(t)c^2 & 0 & 0 & 0 \\ 0 & a^2(t)P(t) & 0 & 0 \\ 0 & 0 & a^2(t)P(t) & 0 \\ 0 & 0 & 0 & a^2(t)P(t) \end{pmatrix}$$

(Equation (C.14) with $U^\mu = (c, 0, 0, 0)$).

2) A body in free fall in the flat space ($K = 0$) Robertson–Walker metric, (7.7), follows a path parameterised by τ with $\theta = \pi/2$ and $\phi = 0$ constants. Determine the differential equations that $t(\tau)$ and $r(\tau)$ must satisfy for the body's path to be a geodesic. Show that the path $r = const.$, $t = \tau$ is a geodesic for any $a(t)$.

3) Repeat question 2) for the Robertson–Walker metrics with $K = \pm1$. Furthermore, find geodesics with $\dot{r} \neq 0$ (still with θ and ϕ constant)

and show that the radial velocity

$$v \to \begin{cases} c & \text{for } K = 1, \ r \to 1; \\ 0 & \text{for } K = -1, \ r \to \infty, \dot{r} \text{ finite.} \end{cases}$$

4) Show that the Robertson–Walker metric,

$$ds^2 = -c^2 dt^2 + L^2(t) \left(\frac{dr^2}{1 - Kr^2} + r^2(d\theta^2 + \sin^2\theta d\phi^2) \right),$$

can be re-written in the form

$$ds^2 = \eta^2(t') \left\{ -c^2 dt'^2 + (1 - Kr^2)^{-1} dr^2 + r^2(d\theta^2 + \sin^2\theta d\phi^2) \right\},$$

and find an expression for $\eta(t')$ in terms of t and $a(t)$. This shows that any Robertson–Walker metric can be written as a time-dependent dilation of a static metric.

5) Starting from the *5-dimensional* Minkowski space metric

$$ds^2 = -dz_0{}^2 + dz_1{}^2 + dz_2{}^2 + dz_3{}^2 + dz_4{}^2, \quad -\infty < z_\mu < \infty, \quad \mu = 0, \dots, 4,$$

a) Show that the metric induced on the 4-dimensional hyperboloid

$$-z_0^2 + z_1^2 + z_2^2 + z_3^2 + z_4^2 = L^2$$

is

$$ds^2 = -c^2 dt^2 + \exp(2ct/L) \left(dx_1{}^2 + dx_2{}^2 + dx_3{}^2 \right),$$

where the coordinates (ct, x_1, x_2, x_3) are defined by

$$z_0 = a \sinh(ct/L) + \frac{1}{2L} e^{ct/L} (x_1^2 + x_2^2 + x_3^2)$$

$$z_4 = L \cosh(ct/L) - \frac{1}{2L} e^{ct/L} (x_1^2 + x_2^2 + x_3^2)$$

$$z_\alpha = e^{ct/L} x_\alpha, \quad \alpha = 1, \dots, 3$$

(L is a constant).

b) Show that in the coordinates (ct', r, θ, ϕ) given by

$$z_0 = \sqrt{L^2 - r^2} \sinh(ct'/L)$$

$$z_1 = \sqrt{L^2 - r^2} \cosh(ct'/L)$$

$$z_2 = r \sin\theta \cos\phi$$

$$z_3 = r \sin\theta \sin\phi$$

$$z_4 = r \cos\theta$$

(which only cover $z_0 + z_1 > 0$), the metric is

$$ds^2 = -c^2 \left(1 - \frac{r^2}{L^2} \right) dt'^2 + \left(1 - \frac{r^2}{L^2} \right)^{-1} dr^2 + r^2 (d\theta^2 + \sin^2 \theta d\phi^2).$$

6) Determine the geodesic equations for a massive object in free fall in de Sitter space-time by using the coordinates in question 6 of Chapter 6, with $M = 0$. Show that, for radial motion with θ and ϕ constant, the velocity determined by an observer stationary at the origin is

$$v = \frac{dr}{dt} = \frac{1}{k} \left(1 - \frac{r^2}{L^2} \right) \sqrt{k^2 - 1 + \frac{r^2}{L^2}},$$

where k is an integration constant arising from a first integral of the t-equation. What happens for $r \geq L$?

Show that there is a maximum velocity in the range $0 < r < L$ and determine the value of r at which this occurs and the red-shift at that point.

7) Question 4) in Chapter 5 asked you to show that a cosmological constant is equivalent to a fluid with zero enthalpy and negative pressure (tension), $P = -\frac{\Lambda c^4}{8\pi G}$. Calculate the magnitude of this pressure for $\Omega_\Lambda = 0.7$.

8) A positively charged particle called the μ^+-meson has a mass of $106 \, \text{MeV}/c^2$. Above what temperature do you expect $\mu^+ - \mu^-$ pairs to be created out of the vacuum by thermal energy? What time after the Big Bang does this correspond to in the standard model of Cosmology presented here?

Appendix A

Tensors of Type (p, q)

It is useful to categorise tensors according to their properties under coordinate transformations. Generalising (4.3), a rank n tensor $T^{\mu_1 \cdots \mu_p}_{\nu_1 \cdots \nu_q}$, with $n = p + q$, is said to be of type (p, q) if its components transform under a coordinate transformation as

$$T^{\mu_1 \cdots \mu_p}_{\nu_1 \cdots \nu_q}(x|_P) = \frac{\partial x^{\mu_1}}{\partial x^{\rho'_1}} \cdots \frac{\partial x^{\mu_p}}{\partial x^{\rho'_p}} \frac{\partial x^{\lambda'_1}}{\partial x^{\nu_1}} \cdots \frac{\partial x^{\lambda'_q}}{\partial x^{\nu_q}} T'^{\rho'_1 \cdots \rho'_p}_{\lambda'_1 \cdots \lambda'_q}(x'|_P). \tag{A.1}$$

Tensors with all their indices lower are called *covariant tenors*, as generalisations of a covector; tensors with all their indices upper are called *contravariant tensors* and are generalisations of a vector.

Generalising (4.19) and (4.23), the covariant derivative in the ρ-direction is

$$T^{\mu_1 \cdots \mu_p}_{\nu_1 \cdots \nu_q; \rho} = \partial_\rho T^{\mu_1 \cdots \mu_p}_{\nu_1 \cdots \nu_q} + \sum_{r=1}^{p} \Gamma^{\mu_r}_{\lambda \rho} T^{\mu_1 \cdots \mu_{r-1} \lambda \mu_{r+1} \cdots \mu_p}_{\nu_1 \cdots \nu_q}$$

$$- \sum_{s=1}^{q} \Gamma^{\lambda}_{\nu_s \rho} T^{\mu_1 \cdots \mu_p}_{\nu_1 \cdots \nu_{s-1} \lambda \nu_{s+1} \cdots \nu_q}. \tag{A.2}$$

One can also generalise Lie derivatives to tensors of arbitrary rank. Again contravariant and covariant indices require a change in sign and (4.8) generalises to

$$(\mathcal{L}_{\vec{V}} T)^{\mu_1 \cdots \mu_p}_{\nu_1 \cdots \nu_q} = V^\rho \partial_\rho T^{\mu_1 \cdots \mu_p}_{\nu_1 \cdots \nu_q} - \sum_{r=1}^{p} (\partial_\rho V^{\mu_r}) T^{\mu_1 \cdots \mu_{r-1} \rho \mu_{r+1} \cdots \mu_p}_{\nu_1 \cdots \nu_q}$$

$$+ \sum_{s=1}^{q} (\partial_{\nu_s} V^\rho) T^{\mu_1 \cdots \mu_p}_{\nu_1 \cdots \nu_{s-1} \rho \nu_{s+1} \cdots \nu_q}. \tag{A.3}$$

Appendix B

The Riemann Tensor

In this appendix, the form of the Riemann tensor will be derived. The basic idea is to use parallel transport as described in §4.6.

Take a vector $\vec{W}$ at a point P and parallel transport it a small distance away from P in the direction of a vector $\vec{U}$. Under a small displacement through $\epsilon\vec{U}$, with $\epsilon \ll 1$, $\vec{W}$ will change to

$$\vec{W} \;\rightarrow\; \left(1 + \epsilon\nabla_{\vec{U}} + \frac{1}{2}\epsilon^2\nabla^2_{\vec{U}}\right)\vec{W} + O(\epsilon^3).$$

Now let's take another unit vector which is not parallel to $\vec{U}$, $\vec{V}$ say, and subsequently make an infinitesimal parallel displacement through $\epsilon\vec{V}$. Under these two shifts,

$$\vec{W} \rightarrow (1 + \epsilon\nabla_{\vec{V}} + \frac{1}{2}\epsilon^2\nabla^2_{\vec{V}})(1 + \epsilon\nabla_{\vec{U}} + \frac{1}{2}\epsilon^2\nabla^2_{\vec{U}})\vec{W} + O(\epsilon^3)$$

$$= \left\{1 + \epsilon(\nabla_{\vec{V}} + \nabla_{\vec{U}}) + \epsilon^2\left(\frac{1}{2}\nabla^2_{\vec{V}} + \frac{1}{2}\nabla^2_{\vec{U}} + \nabla_{\vec{V}}\nabla_{\vec{U}}\right)\right\}\vec{W} + O(\epsilon^3)$$

$$= \left\{1 + \epsilon(\nabla_{\vec{V}} + \nabla_{\vec{U}}) + \frac{1}{2}\epsilon^2(\nabla_{\vec{V}} + \nabla_{\vec{U}})^2 + \frac{1}{2}\epsilon^2\left(\nabla_{\vec{V}}\nabla_{\vec{U}} - \nabla_{\vec{U}}\nabla_{\vec{V}}\right)\right\}\vec{W} + O(\epsilon^3).$$

Had we done this in the opposite order, first parallel transporting in the $\vec{V}$ direction and then in the $\vec{U}$ direction, we would get

$$\vec{W} \rightarrow \left\{1 + \epsilon(\nabla_{\vec{U}} + \nabla_{\vec{V}}) + \frac{1}{2}\epsilon^2(\nabla_{\vec{U}} + \nabla_{\vec{V}})^2 + \frac{1}{2}\epsilon^2\left(\nabla_{\vec{U}}\nabla_{\vec{V}} - \nabla_{\vec{V}}\nabla_{\vec{U}}\right)\right\}\vec{W} + O(\epsilon^3).$$

The difference between these two results is second order in ϵ,

$$\epsilon^2[\nabla_{\vec{V}}, \nabla_{\vec{U}}]\vec{W},$$

where

$$[\nabla_{\vec{V}}, \nabla_{\vec{U}}] = \nabla_{\vec{V}}\nabla_{\vec{U}} - \nabla_{\vec{U}}\nabla_{\vec{V}} \tag{B.1}$$

is called the *commutator* (more precisely, the covariant commutator) of $\nabla_{\vec{V}}$ with $\nabla_{\vec{U}}$. In a curved space we shall see that this commutator is in general non-zero,

$$\nabla_{\vec{V}}\nabla_{\vec{U}} \neq \nabla_{\vec{U}}\nabla_{\vec{V}}.$$

This is how we shall characterise curvature.[1] If we go a small distance in the direction $\vec{U}$ and then a small distance in the direction $\vec{V}$, we get a different result from that which we would get by going the same distance in the direction $\vec{V}$ and then the same distance in the direction $\vec{U}$. The covariant derivatives $\nabla_{\vec{V}}$ and $\nabla_{\vec{U}}$ do not commute.

What we have done here is to take $\vec{W}$ around two sides of a parallelogram in different directions, and we found that the results differ. Alternatively, we can go all the way around the parallelogram and come back to the starting point: go in the $\vec{U}$ direction first, then the $\vec{V}$ direction, then go *back* in the direction $-\vec{U}$ and then back in the direction $-\vec{V}$. The result is

$$\vec{W} \to \left(1 - \epsilon\nabla_{\vec{V}} + \frac{1}{2}\epsilon^2\nabla^2_{\vec{V}}\right)\left(1 - \epsilon\nabla_{\vec{U}} + \frac{1}{2}\epsilon^2\nabla^2_{\vec{U}}\right)$$

$$\times \left(1 + \epsilon\nabla_{\vec{V}} + \frac{1}{2}\epsilon^2\nabla^2_{\vec{V}}\right)\left(1 + \epsilon\nabla_{\vec{U}} + \frac{1}{2}\epsilon^2\nabla^2_{\vec{U}}\right)\vec{W} + O(\epsilon^3)$$

$$= \left\{1 - \epsilon(\nabla_{\vec{V}} + \nabla_{\vec{U}}) + \frac{1}{2}\epsilon^2(\nabla_{\vec{V}} + \nabla_{\vec{U}})^2 + \frac{1}{2}\epsilon^2[\nabla_{\vec{V}}, \nabla_{\vec{U}}])\right\}$$

$$\times \left\{1 + \epsilon(\nabla_{\vec{V}} + \nabla_{\vec{U}}) + \frac{1}{2}\epsilon^2(\nabla_{\vec{V}} + \nabla_{\vec{U}})^2 + \frac{1}{2}\epsilon^2[\nabla_{\vec{V}}, \nabla_{\vec{U}}])\right\}\vec{W}$$

$$= (1 + \epsilon^2[\nabla_{\vec{V}}, \nabla_{\vec{U}}])\vec{W} + O(\epsilon^3).$$

The vector $\vec{W}$ has been parallel transported around a small parallelogram and rotated in space (or Lorentz transformed in space-time); see Figure 4.6,

$$\vec{W} \to \vec{W} + \epsilon^2[\nabla_{\vec{V}}, \nabla_{\vec{U}}]\vec{W} + O(\epsilon^3). \tag{B.2}$$

We shall calculate the resulting change,

$$\Delta_\epsilon\vec{W} = \epsilon^2[\nabla_{\vec{V}}, \nabla_{\vec{U}}]\vec{W} + O(\epsilon^3), \tag{B.3}$$

ignoring terms of order ϵ^3. Focusing on the $O(\epsilon^2)$ term let

$$\Delta\vec{W} := \lim_{\epsilon \to 0} \frac{\Delta_\epsilon\vec{W}}{\epsilon^2} = [\nabla_{\vec{V}}, \nabla_{\vec{U}}]\vec{W}.$$

---

[1] Beware of a factor of $\frac{1}{2}$ here; the commutator of two derivatives is written $[\nabla_{\vec{V}}, \nabla_{\vec{U}}] = \nabla_{\vec{V}}\nabla_{\vec{U}} - \nabla_{\vec{U}}\nabla_{\vec{V}}$, but the anti-symmetrisation of two indices has a factor of one-half: for example, $F_{[\nu\rho]} = \frac{1}{2}(F_{\nu\rho} - F_{\rho\nu})$. This is a standard convention.

It is shown in what follows that this has components

$$\left([\nabla_{\vec{V}}, \nabla_{\vec{U}}] \vec{W} \right)^{\rho} = [\vec{V}, \vec{U}]^{\lambda} W^{\rho}{}_{;\lambda} + R^{\rho}{}_{\lambda\mu\nu} V^{\mu} U^{\nu} W^{\lambda} \qquad (B.4)$$

where

$$R^{\rho}{}_{\lambda\mu\nu} = -R^{\rho}{}_{\lambda\nu\mu} = \partial_{\mu}\Gamma^{\rho}{}_{\lambda\nu} - \partial_{\nu}\Gamma^{\rho}{}_{\lambda\mu} + \Gamma^{\rho}{}_{\sigma\mu}\Gamma^{\sigma}{}_{\lambda\nu} - \Gamma^{\rho}{}_{\sigma\nu}\Gamma^{\sigma}{}_{\lambda\mu} \qquad (B.5)$$

is a rank-4 tensor of type $(1,3)$ called the *Riemann tensor* and

$$[\vec{V}, \vec{U}]^{\lambda} = V^{\rho}\partial_{\rho}U^{\lambda} - U^{\rho}\partial_{\rho}V^{\lambda} \qquad (B.6)$$

is the *Lie commutator* of $\vec{V}$ with $\vec{U}$ – the Lie derivative of the vector $\vec{U}$ in the direction of $\vec{V}$ (see Appendix A).

The first term on the right-hand side of (B.4) represents a parallel displacement of $\vec{W}$ in the direction $\vec{Z} = [\vec{V}, \vec{U}]$, so $\vec{W}$ does not quite come back to the starting point when $[\vec{V}, \vec{U}] \neq 0$. This can be avoided by choosing $\vec{U}$ and $\vec{V}$ to be coordinate basis vectors themselves, for example, $\vec{U} = \frac{\partial}{\partial x^{\mu}}$ and $\vec{V} = \frac{\partial}{\partial x^{\nu}}$, because their components are just constants in this basis and

$$\left[\frac{\partial}{\partial x^{\mu}}, \frac{\partial}{\partial x^{\nu}} \right] = 0, \qquad (B.7)$$

so (B.6) vanishes. But it should be borne in mind that it does not vanish for general $\vec{U}$ and $\vec{V}$.

The second term on the right-hand side of (B.4) is a purely algebraic linear transformation of $\vec{W}$. If $\vec{V}$ and $\vec{U}$ are unit vectors and the connection is metric compatible so that parallel transport preserves lengths, this linear transformation of $\vec{W}$ can only be a rotation, a rotation in the plane defined by the two unit vectors $\vec{U}$ and $\vec{V}$. (In space-time it could be a Lorentz transformation rather than a rotation.) The actual rotation depends on the components of the Riemann tensor and is proportional to ϵ^{2} (B.3), which in turn is proportional to the area of the parallelogram.

To derive (B.4) and (B.5) consider the components of $\Delta\vec{W}$ for general vectors $\vec{V}$ and $\vec{U}$:

$$(\Delta\vec{W})^{\rho} = \left([\nabla_{\vec{V}}, \nabla_{\vec{U}}]\vec{W} \right)^{\rho} = \left([(V^{\mu}\nabla_{\mu}), (U^{\nu}\nabla_{\nu})]\vec{W} \right)^{\rho} \qquad (B.8)$$

$$= V^{\mu}U^{\nu}\left([\nabla_{\mu}, \nabla_{\nu}]\vec{W} \right)^{\rho} + V^{\mu}\left(U^{\nu}{}_{;\mu} \right)W^{\rho}{}_{;\nu} - U^{\nu}\left(V^{\mu}{}_{;\nu} \right)W^{\rho}{}_{;\mu}.$$

Now from (4.23)

$$W^{\rho}{}_{;\nu} = \partial_{\nu}W^{\rho} + \Gamma^{\rho}{}_{\lambda\nu}W^{\lambda}, \qquad (B.9)$$

and we will need $\left(\nabla_\mu \nabla_\nu \vec{W}\right)^\rho = W^\rho{}_{;\nu;\mu}$ to calculate $\Delta\vec{W}$ – the intermediate steps are a bit messy and we shall try to spell them out carefully. We use the formula (A.2) for the covariant derivative to calculate this,

$$W^\rho{}_{;\nu;\mu} = \partial_\mu\left(W^\rho{}_{;\nu}\right) - \Gamma^\lambda_{\nu\mu}\left(W^\rho{}_{;\lambda}\right) + \Gamma^\rho_{\lambda\mu}\left(W^\lambda{}_{;\nu}\right).$$

Using (B.9), this is

$$
\begin{aligned}
W^\rho{}_{;\nu;\mu} &= \partial_\mu\left(\partial_\nu W^\rho + \Gamma^\rho_{\lambda\nu}W^\lambda\right) - \Gamma^\lambda_{\nu\mu}\left(\partial_\lambda W^\rho + \Gamma^\rho_{\sigma\lambda}W^\sigma\right) \\
&\quad + \Gamma^\rho_{\lambda\mu}\left(\partial_\nu W^\lambda + \Gamma^\lambda_{\sigma\nu}W^\sigma\right) \\
&= \partial_\mu\partial_\nu W^\rho + \left(\partial_\mu\Gamma^\rho_{\lambda\nu}\right)W^\lambda + \Gamma^\rho_{\lambda\nu}\partial_\mu W^\lambda - \Gamma^\lambda_{\nu\mu}\partial_\lambda W^\rho \\
&\quad - \Gamma^\lambda_{\nu\mu}\Gamma^\rho_{\sigma\lambda}W^\sigma + \Gamma^\rho_{\lambda\mu}\partial_\nu W^\lambda + \Gamma^\rho_{\lambda\mu}\Gamma^\lambda_{\sigma\nu}W^\sigma
\end{aligned}
$$

and, re-arranging the terms and the indices,

$$
\begin{aligned}
W^\rho{}_{;\nu;\mu} &= \partial_\mu\partial_\nu W^\rho + \Gamma^\rho_{\lambda\nu}\partial_\mu W^\lambda + \Gamma^\rho_{\lambda\mu}\partial_\nu W^\lambda - \Gamma^\lambda_{\nu\mu}\left(\partial_\lambda W^\rho + \Gamma^\rho_{\lambda\sigma}W^\sigma\right) \\
&\quad + \left(\partial_\mu\Gamma^\rho_{\lambda\nu}\right)W^\lambda + \Gamma^\rho_{\sigma\mu}\Gamma^\sigma_{\lambda\nu}W^\lambda.
\end{aligned}
$$

For $\Delta\vec{W}$ we need a commutator, so subtract $W^\rho{}_{;\mu;\nu}$ from this and note that the first four terms on the right-hand side of the preceding expression are symmetric under interchange of μ and ν, and so will vanish in the commutator,

$$W^\rho{}_{;\nu;\mu} - W^\rho{}_{;\mu;\nu} = \left(\partial_\mu\Gamma^\rho_{\lambda\nu}\right)W^\lambda - \left(\partial_\nu\Gamma^\rho_{\lambda\mu}\right)W^\lambda + \Gamma^\rho_{\sigma\mu}\Gamma^\sigma_{\lambda\nu}W^\lambda - \Gamma^\rho_{\sigma\nu}\Gamma^\sigma_{\lambda\mu}W^\lambda \tag{B.10}$$

$$= R^\rho{}_{\lambda\mu\nu}W^\lambda, \tag{B.11}$$

since $[\partial_\mu, \partial_\nu] = 0$ and $\Gamma^\lambda_{\mu\nu} = \Gamma^\lambda_{\nu\mu}$ (the zero torsion condition), with the Riemann tensor given in (B.5).

To summarise, for any contravariant vector field $\vec{W}$,

$$
\boxed{
\begin{aligned}
W^\rho{}_{;\nu;\mu} - W^\rho{}_{;\mu;\nu} &= R^\rho{}_{\lambda\mu\nu}W^\lambda \quad \text{with} \\
R^\rho{}_{\lambda\mu\nu} &= \partial_\mu\Gamma^\rho_{\lambda\nu} - \partial_\nu\Gamma^\rho_{\lambda\mu} + \Gamma^\rho_{\sigma\mu}\Gamma^\sigma_{\lambda\nu} - \Gamma^\rho_{\sigma\nu}\Gamma^\sigma_{\lambda\mu}.
\end{aligned}
}
$$

This extends to tensors of different rank by acting on each index separately; thus, for tensors of type $(2,0)$,

$$T^{\rho\lambda}{}_{;\nu;\mu} - T^{\rho\lambda}{}_{;\mu;\nu} = R^\rho{}_{\sigma\mu\nu}T^{\sigma\lambda} + R^\lambda{}_{\sigma\mu\nu}T^{\rho\sigma}, \tag{B.12}$$

on co-vectors,

$$W_{\rho;\nu;\mu} - W_{\rho;\mu;\nu} = R_\rho{}^\lambda{}_{\mu\nu}W_\lambda, \tag{B.13}$$

on tensors of type $(1,1)$,

$$T^\rho{}_{\lambda;\nu;\mu} - T^\rho{}_{\lambda;\mu;\nu} = R^\rho{}_{\sigma\mu\nu}T^\sigma{}_\lambda + R_\lambda{}^\sigma{}_{\mu\nu}T^\rho{}_\sigma, \tag{B.14}$$

and on tensors of type $(0,2)$,

$$T_{\rho\lambda;\nu;\mu} - T_{\rho\lambda;\mu;\nu} = R_\rho{}^\sigma{}_{\mu\nu}T_{\sigma\lambda} + R_\lambda{}^\sigma{}_{\mu\nu}T_{\rho\sigma}. \tag{B.15}$$

B.1 Algebraic Properties of $R^\mu{}_{\nu\rho\sigma}$

We now derive some properties of the Riemann tensor that are stated in §4.6 without proof. The Riemann tensor defined in (B.5) is of type $(1,3)$, but some symmetries among the indices are more evident if we lower the first index to form a tensor of type $(0,4)$ (see Appendix A),

$$R_{\mu\nu\sigma\rho} = g_{\mu\lambda}R^\lambda{}_{\nu\sigma\rho} = g_{\mu\lambda}\Big(\partial_\sigma\Gamma^\lambda_{\nu\rho} - \partial_\rho\Gamma^\lambda_{\nu\sigma} + \Gamma^\lambda_{\tau\sigma}\Gamma^\tau_{\nu\rho} - \Gamma^\lambda_{\tau\rho}\Gamma^\tau_{\nu\sigma}\Big). \tag{B.16}$$

Now using (4.31),

$$g_{\mu\lambda}\partial_\sigma\Gamma^\lambda_{\nu\rho} = \partial_\sigma\big(g_{\mu\lambda}\Gamma^\lambda_{\nu\rho}\big) - \big(\partial_\sigma g_{\mu\lambda}\big)\Gamma^\lambda_{\nu\rho}$$
$$= \partial_\sigma\big(\partial_\nu g_{\mu\rho} + \partial_\rho g_{\mu\nu} - \partial_\mu g_{\nu\rho}\big) - \big(\partial_\sigma g_{\mu\lambda}\big)\Gamma^\lambda_{\nu\rho},$$

and metric compatibility of the connection demands that the metric is covariantly constant:

$$g_{\mu\lambda;\sigma} = \partial_\sigma g_{\mu\lambda} - \Gamma^\tau_{\mu\sigma}\, g_{\tau\lambda} - \Gamma^\tau_{\lambda\sigma}\, g_{\mu\tau} = 0$$
$$\Rightarrow \quad \partial_\sigma g_{\mu\lambda} = \Gamma^\tau_{\mu\sigma}\, g_{\tau\lambda} + \Gamma^\tau_{\lambda\sigma}\, g_{\mu\tau}.$$

Hence

$$g_{\mu\lambda}\partial_\sigma\Gamma^\lambda_{\nu\rho} = \big(\partial_\sigma\partial_\nu g_{\mu\rho} + \partial_\sigma\partial_\rho g_{\mu\nu} - \partial_\sigma\partial_\mu g_{\nu\rho}\big) - \big(\Gamma^\tau_{\mu\sigma}\, g_{\tau\lambda} + \Gamma^\tau_{\lambda\sigma}\, g_{\mu\tau}\big)\Gamma^\lambda_{\nu\rho}. \tag{B.17}$$

Anti-symmetrising on ρ and σ, we can use (B.17) in (B.16) to arrive at the following expression for the Riemann tensor as a tensor of type $(0,4)$:

$$R_{\mu\nu\sigma\rho}=\frac{1}{2}\big(\partial_\sigma\partial_\nu g_{\mu\rho}-\partial_\sigma\partial_\mu g_{\nu\rho}-\partial_\rho\partial_\nu g_{\mu\sigma}+\partial_\rho\partial_\mu g_{\nu\sigma}\big)+g_{\lambda\tau}\big(\Gamma^\lambda_{\mu\rho}\Gamma^\tau_{\nu\sigma}-\Gamma^\lambda_{\mu\sigma}\Gamma^\tau_{\nu\rho}\big). \tag{B.18}$$

In this form we can easily derive the following properties:

1. By construction, the Riemann tensor is anti-symmetric in the last two indices, but it is also anti-symmetric on the first two indices:

$$R_{\mu\nu\rho\sigma} = -R_{\mu\nu\sigma\rho} = -R_{\nu\mu\rho\sigma}. \tag{B.19}$$

2. $R_{\mu\nu\rho\sigma}$ is symmetric under interchange of the first and second pairs of indices:

$$R_{\mu\nu\rho\sigma} = R_{\rho\sigma\mu\nu}. \tag{B.20}$$

3. If the last three indices are completely anti-symmetrised, the result vanishes,

$$R_{\mu[\nu\rho\sigma]} = 0, \tag{B.21}$$

where $[\nu\rho\sigma]$ indicates the sum of all six permutations of the three indices, with a plus sign for even permutations and a minus sign for odd permutations. This is known as a Bianchi identity, the first of two.

Property 2 is a direct consequence of metric compatibility of the connection, $g_{\mu\lambda;\sigma} = 0$. Applying (B.15) to $g_{\rho\sigma}$,

$$0 = g_{\rho\sigma;\nu;\mu} - g_{\rho\sigma;\mu;\nu} = R^{\lambda}{}_{\rho\mu\nu}g_{\lambda\sigma} + R^{\lambda}{}_{\sigma\mu\nu}g_{\rho\lambda} = R_{\sigma\rho\mu\nu} + R_{\rho\sigma\mu\nu}$$

$$\Rightarrow \quad R_{\sigma\rho\mu\nu} = -R_{\rho\sigma\mu\nu}.$$

B.2 Differential Properties of $R^{\mu}{}_{\nu\rho\sigma}$

By differentiating (B.11) and taking appropriate combinations, we can derive an identity that the covariant derivative of $R^{\mu}{}_{\nu\rho\sigma}$ must satisfy, called the *second Bianchi identity*. From (B.11),

$$\left(\nabla_{\rho}[\nabla_{\mu}, \nabla_{\nu}]\vec{W}\right)^{\lambda} = (W^{\rho}{}_{;\nu;\mu} - W^{\lambda}{}_{;\mu;\nu})_{;\rho}$$
$$= \left(R^{\lambda}{}_{\sigma\mu\nu}W^{\sigma}\right)_{;\rho}$$
$$= \left(R^{\lambda}{}_{\sigma\mu\nu;\rho}\right)W^{\sigma} + R^{\lambda}{}_{\sigma\mu\nu}\left(W^{\sigma}{}_{;\rho}\right).$$

On the other hand, $\left(\nabla_{\rho}\vec{W}\right)^{\lambda} = W^{\lambda}{}_{;\rho}$ is itself a tensor of type $(1,1)$, so, using (B.14),

$$\left([\nabla_{\mu}, \nabla_{\nu}]\nabla_{\rho}\vec{W}\right)^{\lambda} = R^{\lambda}{}_{\sigma\mu\nu}W^{\sigma}{}_{;\rho} + R^{\sigma}{}_{\rho\mu\nu}W^{\lambda}{}_{;\sigma}.$$

Subtracting these, we form the double commutator

$$\left([\nabla_{\rho}, [\nabla_{\mu}, \nabla_{\nu}]]\vec{W}\right)^{\lambda} = \left(\nabla_{\rho}[\nabla_{\mu}, \nabla_{\nu}]\vec{W}\right)^{\lambda} - \left([\nabla_{\mu}, \nabla_{\nu}]\nabla_{\rho}\vec{W}\right)^{\lambda}$$
$$= \left(R^{\lambda}{}_{\sigma\mu\nu;\rho}\right)W^{\sigma} + R^{\sigma}{}_{\rho\mu\nu}W^{\lambda}{}_{;\sigma}. \tag{B.22}$$

Double commutators like this necessarily satisfy an identity called the *Jacobi identity*:

$$[\nabla_{\rho}, [\nabla_{\mu}, \nabla_{\nu}]] + [\nabla_{\mu}, [\nabla_{\nu}, \nabla_{\rho}]] + [\nabla_{\nu}, [\nabla_{\rho}, \nabla_{\mu}]] = 0. \tag{B.23}$$

The Jacobi identity is easily checked by expanding all the commutators in (B.23) and writing out all 12 terms explicitly (keeping careful track of the order of the indices) – everything cancels. If we take this combination and use (B.22), all the covariant derivatives of W^λ on the right-hand side cancel and, using the first Bianchi identity (B.21), we are left with

$$0 = \left([\nabla_\rho, [\nabla_\mu, \nabla_\nu]]\vec{W}\right)^\lambda + \left([\nabla_\mu, [\nabla_\nu, \nabla_\rho]]\vec{W}\right)^\lambda + \left([\nabla_\nu, [\nabla_\rho, \nabla_\mu]]\vec{W}\right)^\lambda$$
$$= \left(R^\lambda{}_{\sigma\mu\nu;\rho}\right)W^\sigma + \left(R^\lambda{}_{\sigma\nu\rho;\mu}\right)W^\sigma + \left(R^\lambda{}_{\sigma\rho\mu;\nu}\right)W^\sigma. \tag{B.24}$$

Since this must be true for any vector field W^λ, it must be the case that

$$R^\lambda{}_{\sigma\mu\nu;\rho} + R^\lambda{}_{\sigma\nu\rho;\mu} + R^\lambda{}_{\sigma\rho\mu;\nu} = R^\lambda{}_{\sigma[\mu\nu;\rho]} = 0. \tag{B.25}$$

This is the second Bianchi identity.

The second Bianchi identity can also be confirmed by brute force, without reference to any vector field $\vec{W}$. First, express the Riemann tensor as a tensor of type $(2,2)$,

$$R^{\lambda\sigma}{}_{\mu\nu} = g^{\sigma\tau} R^\lambda{}_{\tau\mu\nu}.$$

Taking the covariant derivative,

$$R^{\lambda\sigma}{}_{\mu\nu;\rho} = \partial_\rho R^{\lambda\sigma}{}_{\mu\nu} + \Gamma^\lambda{}_{\tau\rho} R^{\tau\sigma}{}_{\mu\nu} + \Gamma^\sigma{}_{\tau\rho} R^{\lambda\tau}{}_{\mu\nu} - \Gamma^\tau{}_{\rho\mu} R^{\lambda\sigma}{}_{\tau\nu} - \Gamma^\tau{}_{\rho\nu} R^{\lambda\sigma}{}_{\mu\tau}, \tag{B.26}$$

it is straightforward, but tedious, to add five extra terms and show that

$$R^{\lambda\sigma}{}_{[\mu\nu;\rho]} = 0. \tag{B.27}$$

In fact, the first Bianchi identity (B.21) can be obtained by applying exactly the same analysis to a function Φ rather than a vector field:

$$\nabla_\nu \Phi = \partial_\nu \Phi, \qquad \nabla_\mu \nabla_\nu \Phi = \partial_\mu \partial_\nu \Phi - \Gamma^\lambda{}_{\mu\nu} \nabla_\lambda \Phi \quad \Rightarrow \quad \nabla_{[\mu} \nabla_{\nu]} \Phi = 0,$$

and the equivalent of (B.22) is

$$[\nabla_\rho, [\nabla_\mu, \nabla_\nu]]\Phi = \nabla_\rho [\nabla_\mu, \nabla_\nu]\Phi - [\nabla_\mu, \nabla_\nu]\nabla_\rho \Phi$$
$$= -[\nabla_\mu, \nabla_\nu]\nabla_\rho \Phi = R^\sigma{}_{\rho\mu\nu} \nabla_\sigma \Phi,$$

since $[\nabla_\mu, \nabla_\nu]\Phi = 0$. Applying the Jacobi identity then gives

$$[\nabla_\rho, [\nabla_\mu, \nabla_\nu]]\Phi + [\nabla_\mu, [\nabla_\nu, \nabla_\rho]]\Phi + [\nabla_\nu, [\nabla_\rho, \nabla_\mu]]\Phi = R^\sigma{}_{[\rho\mu\nu]} \nabla_\sigma \Phi = 0.$$

Since this must be true for any differentiable function Φ, we conclude that

$$R^\sigma{}_{[\rho\mu\nu]} = 0, \tag{B.28}$$

which is property 3.

(B.28) is usually referred to as the first Bianchi identity, while (B.27) is the second Bianchi identity.

The Ricci tensor and its trace, the Ricci scalar, are defined by

$$R_{\mu\nu} := R^\rho{}_{\mu\rho\nu} \quad \text{and} \quad R := R^\mu{}_\mu = g^{\mu\nu} R^\rho{}_{\mu\rho\nu}. \tag{B.29}$$

Property 2 of the Riemann tensor (B.20) implies that $R_{\mu\nu}$ is a symmetric matrix.

We can use (B.27) to obtain the covariant derivative of the Ricci tensor. First, write (B.27),

$$R^{\mu\nu}{}_{\rho\sigma;\lambda} + R^{\mu\nu}{}_{\sigma\lambda;\rho} + R^{\mu\nu}{}_{\lambda\rho;\sigma} = 0.$$

Now contract over the indices ν and σ, set $\nu = \sigma$, and sum over them:

$$R^{\mu\nu}{}_{\rho\nu;\lambda} + R^{\mu\nu}{}_{\nu\lambda;\rho} + R^{\mu\nu}{}_{\lambda\rho;\nu} = R^\mu{}_{\rho;\lambda} - R^\mu{}_{\lambda;\rho} + R^{\mu\nu}{}_{\lambda\rho;\nu} = 0.$$

Now contract over μ and λ to get

$$R^\mu{}_{\rho;\mu} - R^\mu{}_{\mu;\rho} + R^\nu{}_{\rho;\nu} = 2R^\mu{}_{\rho;\mu} - R_{,\rho} = 0.$$

Since the Ricci scalar $R^\mu{}_\mu = R$ has no indices, $\partial_\rho R$ is just an ordinary partial derivative. We arrive at

$$R^\mu{}_{\rho;\mu} = \frac{1}{2}\partial_\rho R. \tag{B.30}$$

Defining the *Einstein tensor* $G_{\mu\nu}$ by

$$G_{\mu\nu} = R_{\mu\nu} - \frac{1}{2}g_{\mu\nu}R \tag{B.31}$$

(B.30) then implies that $G^{\mu\nu}$ is covariantly constant in the sense that

$$G^{\nu\mu}{}_{;\mu} = 0. \tag{B.32}$$

In summary, we have the two Bianchi identities and covariant constancy of the Einstein tensor:

$$\boxed{\begin{aligned} R^\mu{}_{[\nu\rho\sigma]} &= 0, \\ R^\lambda{}_{\sigma[\mu\nu;\rho]} &= 0, \\ G^\mu{}_{\nu;\mu} &= 0, \end{aligned}} \quad \begin{aligned} &\text{(B.33a)} \\ &\text{(B.33b)} \\ &\text{(B.33c)} \end{aligned}$$

where

$$G^\mu{}_\nu = R^\mu{}_\nu - \frac{1}{2}\delta^\mu{}_\nu R. \tag{B.34}$$

Lastly, we shall determine the number of independent components of the Riemann tensor in D dimensions. Equation (B.19) implies that

$$R_{\mu\nu\rho\sigma} = R_{[\mu\nu][\rho\sigma]},$$

so in D dimensions we can think of $R_{\mu\nu\rho\sigma}$ as a square matrix with rows labelled by pairs of indices $[\mu\nu]$, $\frac{1}{2}D(D-1)$ pairs, and columns labelled by pairs $[\rho\sigma]$, so $R_{[\mu\nu][\rho\sigma]}$ is a $\frac{1}{2}D(D-1) \times \frac{1}{2}D(D-1)$ matrix. The second property (B.20) implies that $R_{[\mu\nu][\rho\sigma]} = R_{[\mu\sigma][\mu\nu]}$ is a symmetric $\frac{1}{2}D(D-1) \times \frac{1}{2}D(D-1)$ matrix and therefore has

$$\frac{1}{2}\left\{\frac{1}{2}D(D-1)\right\} \times \left\{\frac{1}{2}D(D-1)+1\right\}$$

components. Property 3 (B.21) imposes further restrictions: there are D possibilities for the index μ but, because of the anti-symmetrisation in μ and ν in (B.19), there are only $(D-1)$ possibilities for the index ν; furthermore, because of the anti-symmetrisation in $[\nu\rho\sigma]$, there are then only $(D-2)$ possibilities for ρ and $(D-3)$ for σ. (B.21) therefore imposes $\frac{1}{4!}D(D-1)(D-2)(D-3)$ further conditions. The total number of degrees of freedom in $R_{\mu\nu\rho\sigma}$ in D dimensions is therefore

$$\frac{1}{2}\left\{\frac{1}{2}D(D-1)\right\} \times \left\{\frac{1}{2}D(D-1)+1\right\} - \frac{1}{24}D(D-1)(D-2)(D-3)$$

$$= \frac{1}{12}D^2(D^2-1).$$

In two dimensions there is only one independent component, so there is complete information about the curvature in the Ricci scalar, in three dimensions there are six (there is complete information about the curvature in the Ricci tensor, a symmetric 3×3 matrix), and in four dimensions there are 20 independent components in $R_{\mu\nu\rho\sigma}$. Only in four dimensions and higher do we really need the Riemann tensor; the Ricci scalar suffices to characterise curvature completely in two dimensions and the Ricci tensor in three dimensions.

Appendix C

The Energy-Momentum Tensor

To motivate the definition of the energy-momentum tensor we start with a fluid in three dimensions in Euclidean space, using Cartesian coordinates for simplicity. Suppose we have a fluid made up of identical particles with mass m and density $\rho = mn$ where n is the number of particles per unit volume. If the material is moving with velocity $\mathbf{v}$, each particle has momentum $m\mathbf{v}$, so there is a momentum per unit volume, a momentum density $\mathbf{p} = nm\mathbf{v}$. If $n(t, \mathbf{x})$ is a function of position and time, then the number of particles in a small volume V will vary as a function of time, which must be compensated by particles entering or leaving V through its bounding surface ∂V. Assuming particles are neither created nor destroyed, conservation of mass demands that

$$\frac{d}{dt} \int_V mn(t, \mathbf{x}) dV = - \int_{\partial V} mn(t, \mathbf{x})\mathbf{v}.d\mathbf{S} \tag{C.1}$$

$$\Rightarrow \int_V \frac{\partial}{\partial t}\big(mn(t, \mathbf{x})\big) dV = - \int_V \nabla.\big(mn(t, \mathbf{x})\mathbf{v}\big) dV \tag{C.2}$$

$$\Rightarrow \int_V \left(\frac{\partial \rho}{\partial t} + \nabla.\mathbf{p}\right) dV = 0, \tag{C.3}$$

where $d\mathbf{S}$ is an infinitesimal outward pointing normal to ∂V. This must be true for any volume V, from which we deduce the continuity equation,

$$\frac{\partial \rho}{\partial t} + \nabla.\mathbf{p} = 0. \tag{C.4}$$

In a relativistic context, let m be the rest mass of the particles and n_0 the number density in the rest frame of the fluid. The mass density in the rest frame is then $\rho_0 = mn_0$. If the fluid is moving with velocity v, the rest mass is replaced by $\gamma(v)m$ (with energy $\gamma(v)mc^2$) and the volume shrinks by a factor $1/\gamma(v)$, due to Lorentz–Fitzgerald contraction, so the mass density is $\rho = \gamma^2 mn_0$ and the energy density is ρc^2. Similarly, the relativistic 3-momentum density becomes $\mathbf{p} = \gamma^2 m n_0 \mathbf{v}$. The continuity equation (C.4) can then be written in relativistic notation, with $x^0 = ct$,

$$\partial_0(\rho c^2) + \partial_\alpha(c p^\alpha), \qquad \alpha = 1, 2, 3. \tag{C.5}$$

Note that $\mathbf{p}c$ has the same dimensions as ρc^2, it is not a momentum per unit volume but a momentum per unit area per unit time.

Internal stresses are a form of energy which manifest themselves as pressures and, while for a gas pressure is just one variable, in a viscous fluid or a solid the description of pressure is more involved – a viscous fluid or a solid can sustain tangential pressures, that is, the force on a small area of material need not be normal to the area. (Electromagnetic fields also generate tangential forces.) Consider a small volume of fluid in equilibrium. On a surface element of area δS^α, normal to the x^α direction, the force

$$d\mathbf{f} = \left(P^{\alpha\beta}\hat{\mathbf{x}}_\beta\right)\delta S_\alpha,$$

where $\hat{\mathbf{x}}_\beta$ is a unit vector in the β-direction, need not be normal to δS_α if the medium can maintain tangential forces. A full description of the internal strains in the medium requires nine parameters $P^{\alpha\beta}$ in general. In an isotropic fluid we expect the force to be parallel to the normal and to have the same magnitude in every direction, so, in Cartesian coordinates in flat space, $P^{\alpha\beta} = P\delta^{\alpha\beta}$, where P is the usual pressure, but more generally, in a viscous fluid we need all nine $P^{\alpha\beta}$. The 3×3 matrix is called the *pressure tensor* whose components $P^{\alpha\beta}$ describe the β-component of the force on a small slab of material perpendicular to the α-direction. If there is to be no excess torque on the fluid element, the forces must balance in such a way that

$$P^{\alpha\beta} = P^{\beta\alpha}. \tag{C.6}$$

The total force on a volume V of fluid can be obtained by summing up the inward pointing forces on the boundary ∂V of V,

$$f^\beta \hat{\mathbf{x}}_\beta = -\int_{\partial V}\left(P^{\alpha\beta}\hat{\mathbf{x}}_\beta\right)dS_\alpha = -\hat{\mathbf{x}}_\beta\int_{\partial V}P^{\alpha\beta}dS_\alpha = -\hat{\mathbf{x}}_\beta\int_V\partial_\alpha P^{\alpha\beta}dV. \tag{C.7}$$

(The minus sign is because $d\mathbf{S}$ is defined to be outward pointing.) If there is a net flow of momentum into V, then this will result in a force on the volume. The force in the β-direction will be the rate of change of the β component of the total momentum in the volume V:

$$f^\beta = \frac{d}{dt}\int_V p^\beta dV = \int_V \frac{\partial p^\beta}{\partial t}dV = \int_V \frac{\partial(p^\beta c)}{\partial x^0}dV = \int_V \partial_0(p^\beta c)dV.$$

Comparing this with (C.7), we conclude that

$$\int_V \partial_0(p^\beta c)dV + \int_V \partial_\alpha P^{\alpha\beta} dV = 0,$$

and since this must be true for any volume V,

$$\partial_0(p^\beta c) + \partial_\alpha P^{\alpha\beta} = 0. \qquad (C.8)$$

Conservation of mass (C.5) and conservation of momentum (C.8) can be neatly combined by defining a symmetric 4×4 matrix,

$$T^{\mu\nu} = \begin{pmatrix} \rho c^2 & p^1 c & p^2 c & p^3 c \\ p^1 c & P^{11} & P^{12} & P^{13} \\ p^2 c & P^{21} & P^{22} & P^{23} \\ p^3 c & P^{31} & P^{32} & P^{33} \end{pmatrix}. \qquad (C.9)$$

Then

$$\partial_\mu T^{\mu\nu} = 0 \qquad (C.10)$$

reproduces (C.5) and (C.8). $T^{\mu\nu}$ is sometimes called the *energy-momentum* tensor and sometimes the *stress-energy* tensor. It should perhaps more properly be called the stress-energy-momentum tensor, but that is rather a mouthful.

For a gas with density ρ and pressure P at rest,

$$T^{\mu\nu} = \begin{pmatrix} \rho c^2 & 0 & 0 & 0 \\ 0 & P & 0 & 0 \\ 0 & 0 & P & 0 \\ 0 & 0 & 0 & P \end{pmatrix}. \qquad (C.11)$$

(C.11) is the energy-momentum tensor for a fluid at rest in Minkowski space, in Cartesian coordinates. If the fluid is moving, we can start with (C.11)in the rest frame and transform to a frame in which the fluid is moving. Suppose we have a fluid moving with speed u in the x-direction; if we boost to a coordinate system, also moving with speed u in the x-direction, the fluid will be at rest in that coordinate system and the energy-momentum tensor will have the form (C.11). To find the energy-momentum tensor in the reference frame in which the fluid is moving, we start with (C.11), with coordinates (ct, x, y, z), and boost by u in the negative x-direction to a coordinate system (ct', x', y', z') using the

Lorentz transformation matrix

$$L^{\mu'}{}_{\nu} = \begin{pmatrix} \gamma(u) & \gamma(u)\frac{u}{c} & 0 & 0 \\ \gamma(u)\frac{u}{c} & \gamma(u) & 0 & 0 \\ 0 & 0 & 1 & 0 \\ 0 & 0 & 0 & 1 \end{pmatrix}.$$

So,

$$T^{\mu'\nu'} = L^{\mu'}{}_{\rho} L^{\nu'}{}_{\lambda} T^{\rho\lambda}$$

or

$$T^{\mu'\nu'} = \begin{pmatrix} \gamma & \gamma\frac{u}{c} & 0 & 0 \\ \gamma\frac{u}{c} & \gamma & 0 & 0 \\ 0 & 0 & 1 & 0 \\ 0 & 0 & 0 & 1 \end{pmatrix} \begin{pmatrix} \rho c^2 & 0 & 0 & 0 \\ 0 & P & 0 & 0 \\ 0 & 0 & P & 0 \\ 0 & 0 & 0 & P \end{pmatrix} \begin{pmatrix} \gamma & \gamma\frac{u}{c} & 0 & 0 \\ \gamma\frac{u}{c} & \gamma & 0 & 0 \\ 0 & 0 & 1 & 0 \\ 0 & 0 & 0 & 1 \end{pmatrix}$$

$$= \begin{pmatrix} \gamma^2\left(\rho c^2 + \frac{u^2}{c^2}P\right) & \gamma^2\frac{u}{c}(\rho c^2 + P) & 0 & 0 \\ \gamma^2\frac{u}{c}(\rho c^2 + P) & \gamma^2(\rho u^2 + P) & 0 & 0 \\ 0 & 0 & P & 0 \\ 0 & 0 & 0 & P \end{pmatrix}. \qquad (C.12)$$

Unsurprisingly, the mass density contributes to the pressure in the x'-direction when the wind blows, but what is not so obvious is that pressure contributes to the energy density at relativistic speeds, in the primed frame

$$\rho' c^2 = \gamma^2 \left(\rho c^2 + \frac{u^2}{c^2}P \right).$$

As far as $\rho \to \gamma^2 \rho$ is concerned, one factor of γ can be understood as coming from the relativistic mass increase, $m \to \gamma m$, and the other comes from Lorentz–Fitzgerald contraction in the x-direction, leading to a reduction in volumes $dx\,dy\,dz \to \frac{1}{\gamma}dx\,dy\,dz$, and a subsequent increase in density – but there is also an extra contribution of $\frac{u^2}{c^2}P$ to the energy density.

More generally, for a boost in an arbitrary 3-dimensional direction $\vec{u}$ we can write the energy-momentum tensor as

$$T^{\mu'\nu'} = \left(\rho + \frac{P}{c^2} \right) U^{\mu'} U^{\nu'} + \eta^{\mu'\nu'} P, \qquad (C.13)$$

where $U^{\mu'} = \gamma(u)(c, u_1, u_2, u_3)$ and $\eta_{\mu'\nu'}$ is the Minkowski metric. The expression (C.14) reduces to (C.12) when $u_2 = u_3 = 0$.

If we are in a curved space-time (or even in flat space-time if we choose to use curvilinear coordinates), we just replace the Minkowski metric in this expression with a more general metric with components $g_{\mu\nu}$ (we can drop the primes) and write

$$T^{\mu\nu} = \left(\rho + \frac{P}{c^2}\right) U^\mu U^\nu + g^{\mu\nu} P. \qquad \text{(C.14)}$$

Appendix D

The Schwarzschild Metric

The most general spherically symmetric static metric can be put into the form

$$ds^2 = -f^2(r)c^2dt^2 + g^2(r)dr^2 + h^2(r)(d\theta^2 + \sin^2\theta d\phi^2), \qquad \text{(D.1)}$$

where $f(r)$, $g(r)$, and $h(r)$ are three functions. We can immediately eliminate one of these functions by a coordinate transformation. Let $\tilde{r} = h(r)$: then,

$$ds^2 = -f^2(\tilde{r})c^2dt^2 + g^2(\tilde{r})\left(\frac{dr}{d\tilde{r}}\right)^2 d\tilde{r}^2 + \tilde{r}^2(d\theta^2 + \sin^2\theta d\phi^2), \qquad \text{(D.2)}$$

where

$$\frac{dr}{d\tilde{r}} = \frac{1}{dh/dr}.$$

Now, let $\tilde{g}(\tilde{r}) = g(\tilde{r})\frac{dr}{d\tilde{r}}$ and

$$ds^2 = -f^2(\tilde{r})c^2dt^2 + \tilde{g}^2(\tilde{r})d\tilde{r}^2 + \tilde{r}^2(d\theta^2 + \sin^2\theta d\phi^2), \qquad \text{(D.3)}$$

which is the same form as (D.1) but with $h(r) = \tilde{r}$. Of course, $f(r)$ and $g(r)$ in (D.1) and (D.4) are different functions, but that doesn't matter; they are undetermined functions at the moment and it is just a matter of notation. So we can drop the tildes and write the most general spherically static metric as[1]

$$ds^2 = -f^2(r)c^2dt^2 + g^2(r)dr^2 + r^2(d\theta^2 + \sin^2\theta d\phi^2). \qquad \text{(D.4)}$$

---

[1] We are assuming that t is a time coordinate and r is a space coordinate, so the coefficient of dt^2 is negative and the coefficient of dr^2 is positive. This assumption is not essential and is easily avoided by writing $f^2(r) = A(r)$ and $g^2(r) = B(r)$ and allowing A and B to change sign.

The connection coefficients are

$$\Gamma^r_{rr} = \frac{1}{g}\frac{dg}{dr} = \frac{1}{2g^2}\frac{dg^2}{dr}, \quad \Gamma^r_{00} = \frac{f}{g^2}\frac{df}{dr} = \frac{1}{2g^2}\frac{df^2}{dr},$$

$$\Gamma^0_{0r} = \Gamma^0_{r0} = \frac{1}{f}\frac{df}{dr} = \frac{1}{2f^2}\frac{df^2}{dr},$$

$$\Gamma^r_{\theta\theta} = -\frac{r}{g^2}, \quad \Gamma^r_{\phi\phi} = -\frac{r\sin^2\theta}{g^2}, \quad \Gamma^\theta_{\theta r} = \Gamma^\theta_{r\theta} = \Gamma^\phi_{r\phi} = \Gamma^\phi_{\phi r} = \frac{1}{r},$$

$$\Gamma^\theta_{\phi\phi} = -\sin\theta\cos\theta, \quad \Gamma^\phi_{\phi\theta} = \Gamma^\phi_{\theta\phi} = \cot\theta.$$

($x^0 = ct$ is used as the time coordinate, rather than t; this avoids factors of c^2 in the Riemann tensor components that follow.)

The non-zero components of the Riemann tensor are obtained from

$$R_{0r0r} = fg\left(\frac{f'}{g}\right)',$$

$$R_{0\theta0\theta} = \frac{rff'}{g^2}, \quad R_{0\phi0\phi} = \frac{rff'}{g^2}\sin^2\theta,$$

$$R_{r\theta r\theta} = \frac{rg'}{g}, \quad R_{r\phi r\phi} = \frac{rg'}{g}\sin^2\theta,$$

$$R_{\theta\phi\theta\phi} = \left(\frac{g^2-1}{g^2}\right)r^2\sin^2\theta,$$

together with (B.19) and (B.20). The Ricci tensor is diagonal,

$$R_{00} = \frac{f\left(g\left(2f'+rf''\right)-rf'g'\right)}{rg^3}$$

$$R_{rr} = \frac{rf'g'-rgf''+2fg'}{rfg}$$

$$R_{\theta\theta} = \frac{f\left(rg'+g^3-g\right)-rgf'}{fg^3}$$

$$R_{\phi\phi} = \frac{\left\{f\left(rg'+g^3-g\right)-rgf'\right\}\sin^2\theta}{fg^3},$$

and the Ricci scalar is

$$R = \frac{2\left\{r\left(rf'g'-g\left(2f'+rf''\right)\right)+f\left(2rg'+g^3-g\right)\right\}}{r^2fg^3}.$$

The Einstein tensor is also diagonal:

$$G_{00} = \frac{f^2\left(2rg'+g^3-g\right)}{r^2g^3}$$

$$G_{rr} = \frac{2rf' - fg^2 + f}{r^2 f}$$

$$G_{\theta\theta} = \frac{r\left\{g\left(f' + rf''\right) - (rf)'\,g'\right\}}{fg^3}$$

$$G_{\phi\phi} = \left(\frac{r\left\{g\left(f' + rf''\right) - (rf)'\,g'\right\}}{fg^3}\right)\sin^2\theta.$$

It is straightforward to check that $R_{\mu\nu} = G_{\mu\nu} = 0$ and $R = 0$ for $f^2 = g^{-2} = 1 - \frac{2GM}{rc^2}$, but that the Riemann tensor does not vanish.

Appendix E

Robertson–Walker Space-Time

When 3-dimensional space is flat, the Robertson–Walker line element for expanding (or contracting, depending on the sign of $\dot{a}$) space-time is

$$ds^2 = -c^2dt^2 + a^2(t)\left(dx_1^2 + dx_2^2 + dx_3^2\right) = g_{\mu\nu}dx^\mu dx^\nu. \qquad \text{(E.1)}$$

For a fluid at rest in this space-time the 4-velocity $U^\mu = (U^0, 0, 0, 0)$ satisfies

$$-c^2 = g_{\mu\nu}U^\mu U^\nu = -(U^0)^2 \quad \Rightarrow \quad U^0 = c.$$

The energy-momentum tensor (C.14) is

$$T^{\mu\nu} = \begin{pmatrix} \rho c^2 & 0 \\ 0 & \frac{1}{a^2(t)}P\delta^{\alpha\beta} \end{pmatrix}$$

and

$$T_{\mu\nu} = \begin{pmatrix} \rho c^2 & 0 \\ 0 & a^2(t)P\delta_{\alpha\beta} \end{pmatrix},$$

since the metric tensor has components

$$g_{00} = -1, \qquad g_{\alpha\beta} = a^2(t)\delta_{\alpha\beta},$$

where $\alpha, \beta = 1, 2, 3$ label spatial coordinates and again $x^0 = ct$. The non-zero connection coefficients arising from (E.1) are

$$\Gamma^0_{\alpha\beta} = \frac{1}{c}a\dot{a}\,\delta_{\alpha\beta}, \qquad \Gamma^\alpha_{0\beta} = \frac{1}{c}\frac{\dot{a}}{a}\delta^\alpha{}_\beta. \qquad \text{(E.2)}$$

The non-zero components of the Riemann tensor are

$$R_{0\alpha0\beta} = -\frac{1}{c^2}a\ddot{a}\delta_{\alpha\beta}, \qquad R_{\alpha\beta\gamma\delta} = \frac{(\dot{a}a)^2}{c^2}\left(\delta_{\alpha\gamma}\delta_{\beta\delta} - \delta_{\alpha\delta}\delta_{\beta\gamma}\right). \qquad \text{(E.3)}$$

The Ricci tensor has components

$$R_{00} = -\frac{3}{c^2}\frac{\ddot{a}}{a}, \qquad R_{\alpha\beta} = \frac{1}{c^2}(\ddot{a}a + 2\dot{a}^2)\delta_{\alpha\beta} \qquad (E.4)$$

and the Ricci scalar is

$$R = \frac{6}{c^2 a^2}\left(\ddot{a}a + \dot{a}^2\right). \qquad (E.5)$$

The Einstein tensor $G_{\mu\nu} = R_{\mu\nu} - \frac{1}{2}Rg_{\mu\nu}$ therefore has components

$$G_{00} = \frac{3}{c^2}\left(\frac{\dot{a}}{a}\right)^2, \qquad G_{\alpha\beta} = -\frac{1}{c^2}(2\ddot{a}a + \dot{a}^2)\delta_{\alpha\beta}. \qquad (E.6)$$

More generally, homogeneity and isotropy of space allow for spatial curvature, and up to a constant re-scaling of space, there are three possibilities,

$$ds^2 = -c^2 dt^2 + a^2(t)\left(\frac{dr^2}{1 - Kr^2} + r^2(d\theta^2 + \sin^2\theta d\phi^2)\right),$$

with $K = 0, \pm 1$ and metric components

$$g_{\mu\nu} = \begin{pmatrix} -1 & 0 & 0 & 0 \\ 0 & \frac{a^2(t)}{1-Kr^2} & 0 & 0 \\ 0 & 0 & a^2(t)r^2 & 0 \\ 0 & 0 & 0 & a^2(t)r^2\sin^2\theta \end{pmatrix}.$$

The three possible values of K are all compatible with 3-dimensional space (constant t) being homogeneous and isotropic, and they correspond to:

- $K = +1$: space has the geometry of a 3-dimensional sphere of unit radius ($w^2 + x^2 + y^2 + z^2 = 1$ in 4-dimensional Euclidean space) and $0 \le r \le 1$.
- $K = -1$: space has the geometry of a 3-dimensional version of the hyperbolic plane.
- $K = 0$: space is just 3-dimensional Euclidean space, in spherical polar coordinates.

As a mathematical device, it is convenient to define a static 3-dimensional metric

$$\tilde{g}_{\alpha\beta} = \begin{pmatrix} \frac{1}{1-Kr^2} & 0 & 0 \\ 0 & r^2 & 0 \\ 0 & 0 & r^2\sin^2\theta \end{pmatrix}$$

in terms of which

$$g_{\alpha\beta} = a^2(t)\tilde{g}_{\alpha\beta}.$$

The non-zero components of the Riemann tensor are

$$R_{0\alpha0\beta} = -\frac{1}{c^2}a\ddot{a}\tilde{g}_{\alpha\beta}, \qquad R_{\alpha\beta\gamma\delta} = \frac{1}{c^2}(\dot{a}^2 + Kc^2)a^2\left(\tilde{g}_{\alpha\gamma}\tilde{g}_{\beta\delta} - \tilde{g}_{\gamma\delta}\tilde{g}_{\beta k}\right).$$

$$(\text{E.7})$$

The Ricci tensor has components

$$R_{00} = -\frac{3}{c^2}\frac{\ddot{a}}{a}, \qquad R_{\alpha\beta} = \frac{1}{c^2}(\ddot{a}a + 2\dot{a}^2 + 2Kc^2)\tilde{g}_{\alpha\beta}, \qquad (\text{E.8})$$

and the Ricci scalar is

$$R = \frac{6}{c^2a^2}\left(\ddot{a}a + \dot{a}^2 + Kc^2\right).$$

The Einstein tensor has components

$$G_{00} = \frac{3}{c^2}\left(\frac{\dot{a}^2}{a^2} + \frac{Kc^2}{a^2}\right), \qquad G_{\alpha\beta} = -\frac{1}{c^2}(2\ddot{a}a + \dot{a}^2 + Kc^2)\tilde{g}_{\alpha\beta}. \quad (\text{E.9})$$

References

[1] A. Einstein, *The Meaning of Relativity*, Routledge (2003).
A discussion of the ideas behind special and general relativity from the mouth of their creator. Based on lectures delivered at Princeton University in 2021 and first published in 1922.

[2] J. B. Hartle, *Gravity: An Introduction to Einstein's Relativity*, Addison-Wesley (2003).
Adopts a similar approach to that of this text: uses the variational principle to discuss geodesics before introducing tensors, but is more detailed and more advanced.

[3] C. W. Misner, K. S. Thorne, and J. A. Wheeler, *Gravitation*, W. H. Freeman (1973).
Extremely comprehensive tour of general relativity (over 1,000 pages), though the cosmology is now somewhat out of date.

[4] S. W. Weinberg, *Gravitation and Cosmology: Principles and Applications of the General Theory of Relativity*, John Wiley and Sons (1972).
More advanced and detailed, with emphasis on observations and physics, but again the cosmology is now out of date.

[5] R. M. Wald, *General Relativity*, University of Chicago (1984).
Another advanced text, more mathematical than [3] or [4].

[6] S. W. Hawking and G. F. R. Ellis, *The Large Scale Structure of Space-Time*, Cambridge University Press (1973).
Very dense and mathematical but an invaluable resource. Gives rigorous proofs that singularities in space-time are inevitable in Einstein's general theory of relativity.

[7] S. W. Weinberg, *The First Three Minutes: A Modern View of the Origin of the Universe*, Basic Books (1993).
A non-technical, popular account of the Big Bang and nucleosynthesis.

[8] A. Liddle, *An Introduction to Modern Cosmology*, Wiley (2003).
An elementary discussion of cosmology and the Big Bang, more up to date and less mathematical than [4].

[9] A. Liddle and D. Lyth, *Cosmological and Large Scale Structure*, Cambridge University Press (2000).
A detailed account of inflation.

Index

Printed in the United States
by Baker & Taylor Publisher Services

Printed in the United States
by Baker & Taylor Publisher Services